WENDY P. TORI

Stability in Model Populations

MONOGRAPHS IN POPULATION BIOLOGY

EDITED BY SIMON A. LEVIN AND HENRY S. HORN

Titles available in the series (by monograph number)

Stability in Model Populations

LAURENCE D. MUELLER
AND
AMITABH JOSHI

PRINCETON UNIVERSITY PRESS

PRINCETON AND OXFORD

Library of Congress Cataloging-in-Publication Data
Mueller, Laurence D., 1951–
    Stability in model populations / Laurence D. Mueller and Amitabh Joshi.
        p. cm. — (Monographs in population biology ; 31)
    Includes bibliographical references (p.).
    ISBN 0-691-00732-2 (cl : alk. paper) — ISBN 0-691-00733-0 (pbk. : alk. paper)
        1. Population biology. I. Joshi, Amitabh, 1965– II. Title. III. Series.

QH352 .M94 2000
577.8′8—dc21

                                                                00-038493

This book has been composed in Baskerville

The paper used in this publication meets the minimum requirements of
ANSI/NISO Z39.48-1992(R1997) (*Permanence of Paper*)

www.pup.princeton.edu

Printed in the United States of America

10 9 8 7 6 5 4 3 2 1

10 9 8 7 6 5 4 3 2 1
(Pbk.)

To

Carol, Adrienne, and Aellana,

and

Vani and Siddarth

Preface and Acknowledgements

In this book, we have chosen for our focus a very small part of the universe of ecological problems. We have been guided in our efforts by the belief that ecological systems are inherently complicated and that insights will be gained slowly with much work. Nevertheless, we believe that in recent years substantial progress has been made in understanding the forces responsible for determining population stability. Much of this understanding has come about through systematic, long-term studies on laboratory systems, and these form the main subject of our discussion. Those areas in which our knowledge is deepest have been explored successfully, in the best traditions of science, with a combination of theoretical models and experimental tests. An overarching goal of this book is the prominent display of work that has used the inferential power of modeling and the strong inference potential of well-designed experiments. We think ecology can only benefit from continued application of these methods.

The book has benefited from the efforts of many people. Professor Michael B. Usher initially suggested that we develop and expand some of our earlier work in a book format. He also made many constructive suggestions during the early stages of the project. The book has also benefited from the excellent reviews and suggestions of Jim Cushing, Brian Dennis, Bob Desharnais, Steven Ellner, Niranjan Joshi, Vidyanand Nanjundaiah, Timothy Prout, Diego Rodríguez, Michael Rose, Vijay Sharma, Peter Turchin, and Prabhakar Vaidya. The research on *Drosophila* summarized in chapter 6 represents more than 25 years of research. The two of us clearly could not have accomplished it alone. To all our undergraduate students—Dan Borash, Phuc Huynh, Yoshi Morimoto, Jason Shiotsugu, and Vaughn Sweet—we extend our sincere gratitude for all your hard work. A. J. especially thanks V. Sheeba and M. Rajamani for their assistance in carrying out laboratory studies on the dynamics of small populations and metapopulations. Much of this research has been supported by grants from the U.S. National Institutes of Health and National Science Foundation, and from the Department of Science and Technology of the Government of India. We would also like to thank Kuke Bijlsma, Brian Dennis, and their colleagues for permitting us to cite their unpublished work. Sam Ellworthy has been most helpful in the practical aspects of finishing this project and making this book a reality. Finally, we need to thank Francisco Ayala, C. R. Babu, Marcus Feldman, Mike Moody, Timothy Prout, and John Thompson, since our research and view of science is strongly tied to those from whom we have learned.

Contents

CONTENTS

Stability in Model Populations

Introduction

Interest in understanding the growth of biological populations has its origins in the early nineteenth century, predating Haeckel's coining of the word *ökologie* in 1866, denoting the study of the relationship of the organism and its environment, as well as the crystallization of population ecology as a distinct discipline in the 1920s and 1930s. Both Carolus Linnaeus and Thomas Malthus recognized that the very nature of reproduction implies a geometrical increase in population size, and that such an increase obviously does not take place unchecked for most, if not all, biological populations. Charles Darwin, in his 1859 book *The Origin of Species*, drew on this idea of the potentially limitless power of increase of populations. He emphasized that there existed in the natural world a constant struggle for existence that, in turn, formed the framework within which natural selection could act on heritable differences; these differences enabled some individuals to face the vagaries of this struggle better than others.

Two major issues arose directly from this Darwinian view of nature, and the domain of ecology has largely been related to seeking their resolution. The first issue was to understand exactly how the biotic and abiotic environments of a species interact to produce the struggle for existence. The development of physiological ecology in the early 1900s was a direct attempt to study empirically the mechanics of how organisms were able to withstand the rigors of existence in an often hostile world. The second issue—the resolution of which forms the domain of population ecology, and with which we primarily concern ourselves in this book—was the question of how order in nature was maintained in the face of the seemingly complex and disorderly struggle for existence. Clearly, the notion of stability, in the general sense of maintaining an orderly spatial and temporal distribution of living organisms, is implicit in this question. Indeed, as early as 1864, Herbert Spencer, in his book *The Principles of Biology*, argued that the long-term persistence of populations or species implied that forces contributing to mortality were in equilibrium with forces contributing to the preservation of life and to reproduction. He also argued that similar adjustments must exist at the individual level between the organism's ability to sustain its life and its ability to reproduce. Spencer thus foreshadowed by many decades the basic tenets of population ecology and life-history evolution.

Spencer's ideas greatly influenced the development of ecology in the last years of the nineteenth century, shaping the view that nature existed in a balance for the common good, and that the primary job of ecologists was to understand the precise system of checks and balances, through agencies such as competition and predation, that maintained the harmony of nature. More important, several of the major questions that were the subject of principal debates in population ecology during the twentieth century can be seen to arise quite naturally from Spencer's views on the balance of nature. One of the first questions that arises in this context is how exactly one defines the balance of nature, and, having defined it, how one empirically determines whether any given population maintains this balance. Debate on this issue continues, although we have come a long way in clarifying what exactly we mean by notions of population regulation and equilibria (reviewed in Turchin, 1995a). The obvious next issue, once a definition of stability is established, is to understand the proximal causes of stability in biological populations. Relevant studies have examined how exactly populations are regulated to maintain an equilibrium, and what the relative role of biotic and abiotic factors is in population regulation. These studies have

4

been the subject of intense debate in population ecology, especially from the 1930s to the 1950s. Here, too, our understanding has progressed over the decades, with an increasing emphasis now being placed on elucidating the impact of details of the life history of specific organisms—as well the spatial structuring of populations—on the dynamics and stability of populations. And finally, there is the issue of the ultimate causes, if any, for the stability of populations—an issue that explicitly brings evolution into the picture. The focus here is to try to understand how the dynamic behavior and stability characteristics of populations may themselves evolve, perhaps directly by group selection, or indirectly as a by-product of evolutionary changes in life-history characters that are the primary focus of natural selection acting at the level of the individual.

We concern ourselves in this book with all three of the major questions just outlined. One of our primary concerns is to highlight the potentially important role of empirical studies on model systems of laboratory populations in examining both the proximal and ultimate causes of population stability. Indeed, this is our major focus. No doubt, only studies on natural populations can address the issue of whether populations in the real world tend to show stable dynamic behavior. To evaluate the various hypotheses regarding the proximal and ultimate determinants of population dynamics and stability, however, we need to work with systems of replicated populations wherein the degree of environmental complexity and variability can be rigorously controlled and simplified. This is typically possible only with laboratory systems whose basic biology and laboratory ecology are relatively well understood. Indeed, some of the earliest experimental work in population ecology was done on laboratory populations of insects (Pearl, 1927, 1928; Chapman and Baird, 1934) and protozoans (Gause, 1934). Thereafter, experimental work on laboratory systems became less common, especially from the 1960s on, as the emphasis shifted to studies on natural populations. This shift occurred in part because of the ongoing debates over whether natural populations were density regulated, and whether interspecific competition was the primary force shaping the structure of biological communities. This is essentially the current situation: The bulk of research in population ecology now involves either theoretical modeling or the study of natural populations. Both these aspects of population ecology have been the subject of numerous recent books (e.g., Rhodes et al., 1996) and reviews, and we therefore devote less attention to them.

A discussion of what exactly is meant by stability, and how one can empirically assess the stability characteristics of a population, is a necessary prerequisite to dealing with the causes—both proximal and ultimate—of stability in biological populations. We therefore first take up the theory of population stability in chapter 2 and discuss the derivation of stability properties for a range of population growth models, including extremely simple heuristic models as well as more complex ones that take into account various meaningful aspects of the biology of specific systems. In chapter 3 we compare and discuss the respective strengths and weaknesses of several commonly used techniques of assessing population stability. The use of some of these techniques in assessing the stability of natural populations is briefly reviewed in chapter 7, thus addressing the issue of whether populations in nature tend to exhibit stability. Chapters 4 through 6 focus on three model systems that have been the subject of fairly extensive empirical investigations in population ecology: the blowfly *Lucilia cuprina*, flour beetles of the genus *Tribolium*, and the fruitfly *Drosophila melanogaster*. Our goal here is not only to review the results obtained from studies on these model systems but also to make comparisons across systems to see if the diverse results can be explained in the context of some common theoretical framework. We also hope that a detailed examination of these systems serves to highlight some of the advantages of using model laboratory systems to address the question of proximal and ultimate causes of population stability. Indeed, laboratory populations constitute a powerful system in which environmental factors can be varied, one or a few at a time, and in which the consequences of such manipulations on population dynamics can be observed with a high degree of rigor. Laboratory populations of species such as *Drosophila* also hold great promise for investigations on the interface of population ecology and population genetics. Thus, in the concluding chapter, we attempt to put the findings from the three model systems into perspective, as well as to outline what we feel are some of the more interesting unanswered questions in this field. We also discuss ways in which these questions may be addressed in the future by viewing them in the light of a general heuristic framework for understanding the dynamics of stage-structured populations. In this framework, the population dynamic consequences of the density dependence of different life-history stages and fitness components are seen to depend critically on the relationship between two stages: (1) the life stage that is directly controlled by density dependence and (2) the life stage that is the focus of density-dependent regulation of recruitment into either the adult or the juvenile stages.

As a prelude to dealing with the theory of population stability and its applications in the next three chapters, we now outline, in the remainder of this chapter, the historical development of ideas regarding stability in population ecology. We also discuss the various ways in which stability is defined and studied in different contexts.

HISTORICAL DEVELOPMENT OF THE CONCEPT OF POPULATION STABILITY

As we have seen, the notion of population stability as a balance between mortality and reproduction goes back well over 100 years. Indeed, population regulation has been the focus of major debates and discussions in population ecology practically since its inception as a distinct discipline in the early twentieth century. Many of the early arguments about population regulation focused on whether population numbers were controlled primarily by biotic (Howard and Fiske, 1911) or climatic (Uvarov, 1931) factors. In the former case, there was no clear notion of intrinsic density-dependent population regulation by negative feedback, as much of this discussion on biotic factors actually involved predators such as birds that did not impact the prey insect populations in a density-dependent manner. Nicholson (1933) first made the point that "for the production of balance, it is essential that a controlling factor should act more severely against an average individual when the density of animals is high, and less severely when the density is low," thus clearly enunciating the idea of density-dependent regulation of populations. Nicholson (1933, 1954b) also made a clear distinction between "responsive" factors (those affected by population density) and "non-responsive" factors, such as climate or other aspects of the physical environment that do not result in regulation of population density. (The latter may, however, greatly influence the level at which the regulatory mechanisms become operative, and thus may determine the equilibrium size of the regulated population.) Among the responsive factors, Nicholson differentiated between those that were reactive and nonreactive, pointing out that to play a role in regulation, a factor must not only be influenced by population density but must also exert a negative feedback on population density.

Not all ecologists, however, immediately accepted Nicholson's arguments for the primacy of density-dependent factors in population regulation. Andrewartha and Birch (1954) and Den Boer (1968) separately argued that density-dependent factors need not be invoked to explain the regulation of populations, especially the prevention of outbreaks, and that density-independent factors alone could explain the apparent stability of natural populations. Milne (1958), Dempster (1983), and Strong (1986) separately developed the idea of imperfect density dependence, or density vagueness, which essentially said that populations may be only strongly density regulated at fairly high densities, whereas for a large range of intermediate densities population size may fluctuate in a random manner. The consensus of opinion now appears to be that density dependence is a necessary prerequisite for population regulation. Accordingly, the focus of present study has shifted to more detailed analyses of how the life history and ecology of different organisms interact to produce specific patterns of density dependence of different fitness components, and how these patterns of density dependence affect the dynamic behavior of populations (Cappucino and Price, 1995).

WHAT IS STABILITY?

Our frame of reference in this book is on the analysis of single populations. In a deterministic environment, the dynamics of a biological population may be described by a difference or a differential equation. When these equations admit an equilibrium point, it is considered to be stable if perturbations away from this equilibrium result in the system returning to the equilibrium point (Lewontin, 1969). In the next chapter we describe in more detail the idea of local stability. As the name implies, the conclusions are valid only in a small region around the equilibrium point. For some models it is possible to establish whether an equilibrium is globally stable, meaning that the system will converge to the equilibrium point from any feasible starting point. Establishing global stability for experimental systems is typically much more difficult than examining local stability. Consequently, we focus almost entirely on local stability analyses. Any discussion of stability is premised on the assumption that the appropriate time scale and spatial limits of a population are known (Connell and Sousa, 1983). This is usually not a problem for laboratory populations in which the appropriate time and spatial scales are known. For natural populations, however, it is often difficult to unequivocally establish the appropriate time scale and spatial limits.

In any habitat, there is almost always some degree of random variation in the environment that affects the number of organisms present in a particular population. Thus, probably no natural or laboratory population exists in a completely deterministic environment. We expect that random or stochastic variation is a more prominent component of the dynamics of natural populations than of laboratory populations. In fact one reason for bringing populations into the laboratory is to reduce the level of stochastic variation.

The meaning of stability in a stochastic environment is not as clear-cut as in deterministic environments. Turelli (1978) reviews several criteria that might serve as useful measures of stability in random environments: (1) Does the stochastic process describing population dynamics possess a stationary distribution? The answer to this question might be yes if certain conditions are satisfied. A stationary distribution is a probabilistic description of the possible sizes a population may assume. It is "stationary" because it applies regardless of the starting point: No matter what state the population starts out in, the system is expected to converge to this distribution. In this sense, the concept of a stationary distribution is more like the concept of global stability. (2) A stochastic population may also be considered "stable" if the fluctuations about its equilibrium are not too severe. This means the variance or coefficient of variation would need to be less than some bound. This concept is more akin to the traditional concept of local stability mentioned previously, and it would provide a numerical estimate of stability (in the form of the coefficient of variation). Royama (1977, 1992) calls populations having bounded variance and no trend in population size "persistent populations." (3) The expected time to extinction or mean persistence time could also be used as a measure of stability (Ludwig, 1975, 1976). This is a practical property, since there is now great interest in managing endangered populations and assessing the various factors that may lead to extinction.

Extinction times due to demographic variation were studied by MacArthur and Wilson (1967). Their analysis showed that extinction times increase rapidly with increasing carrying capacity. However, an important component of population viability is the frequency of catastrophes that reduce population size (Mangel and Tier, 1993). It is reasonable that these types of rare events, although important, will be related to aspects of the environment rather than the biological properties of density regulation. In this book, our focus is on the biological phenomena that determine population stability rather than on the random aspects of the environment.

Most of the analyses of model populations presented in this book focus on the stability of the deterministic processes that affect population stability. For some populations, especially in nature, the actual dynamics may be quite different than the predictions from the deterministic models. In chapter 3 we discuss in more detail the relative merits of stochastic and deterministic evaluations of stability.

Another type of model stability is sometimes called structural stability. If model assumptions or parameters are changed slightly and the model displays qualitatively new behavior, then the model is structurally unstable. For instance, the neutrally stable cycles predicted by the Lotka-Volterra predator-prey models disappear if the prey growth is assumed to be density dependent or if the predator exhibits satiation. Thus, the exploration of a model's robustness may also reveal its structural stability. We occasionally explore this idea by examining the predictions of several different models. Although the examination of structural stability of models is not common in population biology, there are good examples of this in the literature (Gilpin, 1975, chapter 7).

STABILITY IN METAPOPULATIONS

A major contributor to the high profile of population dynamics studies in ecology in recent years has been the renewed interest in understanding the dynamics of systems of small to moderate-sized populations that are linked by migration (metapopulations). It is becoming clear that a metapopulation view may be of tremendous importance for conservation (Harrison, 1994), biological control (Van der Meijden and van der Veen-van Wijk, 1997), and epidemiology (Earn et al., 1998); it may also provide insights into how natural diversity is structured. Earlier in the twentieth century, Sewall Wright (1931, 1940) pointed out that such evolution could proceed very rapidly in spatially structured populations, especially if the substructuring was accompanied by relatively frequent extinction of local populations and the recolonization of the vacant patches by individuals from neighboring subpopulations. Population ecology, however, remained focused primarily on single populations, although some workers did emphasize the importance of considering spatial structure and local extinction (Andrewartha and Birch, 1954; Huffaker, 1958; Gadgil, 1971). In the early theoretical studies on metapopulation dynamics, emphasis was given to the system as a "population of populations." Hence, the primary focus of these studies was on population turnover and the attainment of a steady state in which some constant proportion of suitable habitat patches was occupied by local populations at any given point in time; the actual proportion occupied depended on the balance between extinction and colonization rates (Levins, 1969, 1970), yielding $\hat{P} = 1 - e/m$, where $\hat{P}$ is the equilibrium proportion of occupied patches, and e and m are extinction and colonization rates, respectively. Implicit in the classical view of metapopulation dynamics was the assumption that local dynamics involve a timescale much smaller than that of the dynamics of extinction and colonization, such that all patches are either empty or fully occupied, and migration does not affect local dynamics (Hanski and Gyllenberg, 1993).

With an increasing realization that incorporating migration into simple population models can have fairly significant effects on local dynamics (McCallum, 1992; Hastings, 1993; Hastings and Higgins, 1994; Sinha and Parthasarathy, 1994, 1996), a somewhat more elaborate view of metapopulation stability at both the global (i.e., metapopulation) and local (i.e., subpopulation) levels is beginning to emerge, (Ruxton, 1994, 1996a; Rohani et al., 1996; Amarasekare, 1998; Doebeli and Ruxton, 1998). The principal differences between this approach and the classical view of metapopulation dynamics are (1) the recognition that the patches in a metapopulation may have different area, suitability, local dynamics, and connectivity to other patches; and (2) the realization that migration rates may be sufficiently high so as to impinge on the local dynamics of subpopulations (Hanski and Simberloff, 1997). A considerable body of theory has now been built up around the interactions of migration rates and local dynamics, and the consequences of this interaction for the stability of the local dynamics as well as for the stability, in terms of total number of individuals rather than proportion of occupied patches, for the metapopulation as a whole. In this section, we briefly discuss some of the predictions arising from this theory. In chapter 6, we describe a recent study in which some of these predictions were tested using experimental laboratory metapopulations of *Drosophila*. Although little empirical work has been done on the interaction of migration rates and stability in metapopulations, we spend some time on this issue because it seems to us that our understanding of local and global stability in metapopulations could be greatly enhanced by work on model laboratory systems. Indeed, a few laboratory studies on the impact of metapopulation structure on extinction of populations have highlighted the utility of laboratory systems for these kinds of investigation (Forney and

Gilpin, 1989; Burkey, 1997). Rigorous field studies testing predictions about the impact of migration on metapopulation dynamics are extremely difficult to conduct, largely as a consequence of the difficulties of empirically establishing the dynamics of local populations and estimating colonization and migration rates in the field (Ims and Yoccoz, 1997; Stacey et al., 1997). It typically requires immense effort even to demonstrate that a particular assemblage of field populations fulfills the criteria for being considered a metapopulation (e.g., Hanski et al., 1994; Harrison and Taylor, 1997; Lewis et al., 1997; Morrison, 1998).

Many of the theoretical studies on metapopulation stability that explicitly incorporate local dynamics were meant to ask whether metapopulation structure (patchiness) could stabilize systems of interacting species (competitors, mutualists, host-parasitoid, or predator-prey systems) that would otherwise result in one or more of the interacting species going extinct (e.g., Levins and Culver, 1971; Hastings and Wolin, 1989; Caswell and Cohen, 1991; Sabelis et al., 1991; Nee and May, 1992; Hanski and Zhang, 1993; Comins and Hassell, 1996). Stability in these models was thus viewed in the sense of ensuring long-term coexistence of the interacting species, and some empirical studies have attempted to test whether migration among patches really helps in ensuring coexistence of interacting species. In some continuous-time predator-prey models, the introduction of spatial heterogeneity may lead to chaos that is otherwise not observed with spatial homogeneity (Pascual and Caswell, 1997). Unfortunately, the evidence from field studies, however carefully conducted, is often of a tentative and qualitative nature. There is evidence that the added spatial dimension of metapopulation structure can ensure coexistence, over periods far longer than a single patch would sustain, of several greenhouse and field systems. These include herbivorous spider mites and their predators (Huffaker, 1958; Laing and Huffaker, 1969; Nachman, 1981, 1991;

van de Klashorst et al., 1992; Walde, 1995); the competing boreal mosses *Tetraplodon angustatus, T. mnioides, Splachnum ampullaceum,* and *S. luteum* (Marino, 1991a, 1991b); and the ragwort *Senecio jacobaea,* its herbivore moth, *Tyria jacobaeae,* and its parasitoid, *Cotesia popularis* (van der Meijden et al., 1991; van der Meijden and van der Veen-van Wijk, 1997). At the same time, herbivore-host plant interactions can often be destabilized by metapopulation structure, especially when interpatch distances are large relative to the dispersal ability of predators or parasitoids of the herbivore (Kareiva, 1987; Roland and Taylor, 1995). Qualitative predictions about extinction versus persistence of interactions, at least within the timescale of empirical studies, are relatively easy to test in the laboratory and in the field. This is can be done either in controlled experiments or through access to long-term records in which one of the species in the interaction is an economically important pest.

The more specific theory dealing with the impact of migration on local and global dynamics in metapopulations is not so easily tested empirically, especially under field conditions. The predictions here are more detailed and also take into account and address the nature of the local dynamics in the individual subpopulations. For example, some models suggest that increasing migration rates tend to increase the coherence among subpopulations exhibiting relatively large fluctuations in numbers. This increase occurs through a synchrony of fluctuations across subpopulations, thus bringing them into phase with each other (McCallum, 1992; Hastings, 1993; Holt and McPeek, 1996; Ranta et al., 1997a; but see also Ruxton, 1996a). This effect could be destabilizing at the metapopulation level as it would cause total metapopulation size to fluctuate with a relatively higher amplitude, raising the likelihood of a chance extinction of the entire assemblage of local populations. Clearly, in this context, global noise associated with large-scale effects such as climatic variations is a correlating influence that tends to synchronize local dynamics (Ranta et al., 1997b; Earn et al., 1998; Grenfell et al., 1998) and can therefore be globally destabilizing if the subpopulations are not relatively stable. Local noise, on the other hand, tends to desynchronize the fluctuations of individual subpopulations. Chaotic local dynamics can magnify the desynchronizing effect of local noise, leading to enhanced stability at the metapopulation level as a result of different subpopulations fluctuating out of phase (Solé and Valls, 1992; Adler, 1993; Allen et al., 1993). Thus, broadly speaking, many models suggest that greater migration in metapopulations is likely to be destabilizing at the global level when local dynamics involve large fluctuations in numbers. However, migration alone, in the absence of global noise, may not be able to enforce synchrony if the local fluctuations are erratic and of large amplitude (Haydon and Steen, 1997).

On the other hand, some models suggest that migration, especially if density dependent, could play a stabilizing role at the metapopulation level by acting to stabilize the local dynamics of subpopulations. In general, migration even at constant rates can stabilize chaotic dynamics of simple population models such as the linear and exponential logistic models by altering the behavior to either sustained periodic cycles or stable equilibria (Sinha and Parthasarathy, 1994; Parthasarathy and Sinha, 1995). Constant immigration or emigration terms can also significantly alter the dynamics of extinction in these simple population models (Sinha and Parthasarathy, 1996). Similarly, density-dependent migration can have a stabilizing local effect by suppressing the fluctuations of individual subpopulations, thus also reducing overall fluctuations in metapopulation size. In fact, in systems of populations in which the local dynamics are chaotic and follow the exponential logistic model, introduction of low levels of migration can actually stabilize local dynamics, with subpopulations exhibiting either stable cycles or stable equilibria rather than chaos (Ruxton, 1994). Yet other theoretical studies suggest that migration in a single-species metapopulation in which local dynamics follow any of a variety of simple discrete-time models may be expected to have a negligible effect on stability at the local level (Hastings, 1991; Gyllenberg et al., 1993; Hassell et al., 1995; Rohani et al., 1996; Ruxton, 1996b).

It is clear from the body of theoretical work on this issue that how exactly migration rates affect local and global stability in metapopulations depends on a multitude of factors; these include the form of local dynamics, the nature (local or global, density dependent or independent) and magnitude of migration among subpopulations, and the extent and magnitude of local and global noise. Many of the details of how these factors interact obviously await further theoretical work. It is also evident that there has been practically no empirical work on the effects of migration on the stability of the dynamics of metapopulations, as opposed to stability in the sense of persistence versus extinction. The most recent comprehensive review of metapopulation biology (Hanski and Gilpin, 1997) fails to mention even one empirical study examining this important issue. One of the reasons for this state of affairs, we feel, is the almost exclusive focus on field studies in metapopulation biology, with a few notable exceptions (e.g., Huffaker, 1958; Forney and Gilpin, 1989; Nachman, 1991; Burkey, 1997). In any empirical test of a model's predictions of the effect of migration on metapopulation dynamics, it is imperative that the experimenters be able to manipulate local dynamics and migration rates at will. This type of manipulation is extremely difficult, if not impossible, to do under field conditions. However, such fine control over the dynamics of real populations can be attained in a laboratory setting, as we discuss in chapters 5 and 6. Thus, laboratory systems may be of great significance in providing the means for empirical validation of more detailed predictions about the interactions among migration, noise, and underlying dynamics in metapopulations. Empirical evidence of these interactions may in turn catalyze the development of more appropriate models of metapopulation dynamics. This field of work is in an embryonic stage at present. Nevertheless, we hope that the foregoing account, as well as our discussion in chapter 6 of an empirical study of laboratory metapopulation dynamics, will draw the attention of readers to the vast potential of laboratory systems in this regard.

WHY ARE WE INTERESTED IN STABILITY?

The stability of populations is intimately related to the factors that determine population growth and are, thus, of obvious interest to ecologists. However, there are several reasons for specifically wanting to understand the general stability properties of populations; some of these reasons are related to major problems in conservation biology and evolutionary biology.

Population Extinction

It seems logical that one consequence of unstable population dynamics would be an increased chance of population extinction. This assumption has led some workers to suggest that populations with unstable dynamics are observed rarely because such populations go extinct at higher rates than do more stable populations (Thomas et al., 1980; Berryman and Millstein, 1989). On the contrary, Allen et al. (1993) suggested that chaotic population dynamics in conjunction with population substructure may enhance species persistence. Although the relationship between population stability and extinction is not simple, the two are clearly intimately related.

Effective Population Size

An important force in the evolution of populations is random genetic drift, and the magnitude of drift a population undergoes is inversely proportional to its size. If the effective population size is reduced, rare genetic variants tend to be lost from the population, and loci are more likely to become homozygous. Moreover, in relatively small populations, selection will be less effective at either increasing or decreasing the frequency of alleles with small effects on fitness. When the size of a population varies over generations, the effective population size is equal to the harmonic mean population size, and is thus especially sensitive to small population sizes. Consequently, a few generations of fairly low numbers can cause a disproportionate decrease in effective population size. Thus, even if a population varies randomly and symmetrically about some mean population size, the effective population size will decline as the amplitude of the fluctuations about the mean size increases (fig. 1.1).

Such variation in population size can be induced either by random variation in environmental factors or by the nature of density-dependent regulatory mechanisms. In both cases,

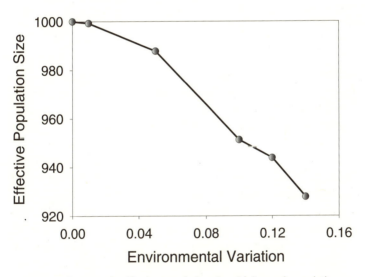

FIG. 1.1. Decrease in effective population size with increasing variation in symmetric random fluctuations about the mean size. Data were generated by simulating 100 generations of population growth with the dynamics governed by a logistic equation with $r = 1.8$ and $K = 1000$. For every generation, a random variable ($\sim N(0, s^2)$) was added to the log of the population size: the value of s is shown on the x-axis.

variation in populations size will tend to reduce the effective population size, rendering the population more susceptible to the effects of random genetic drift.

Fitness in Age-Structured Populations

Fitness in age-structured populations depends on a weighted average of genotypic age-specific survival and fertility values (Charlesworth, 1994). Typically, these fitnesses are computed under the assumption that a population has a stable age distribution. Therefore, if a density-regulated population is not at a stable equilibrium point, these fitness calculations will be wrong. This problem exists even if selection itself is not density dependent. Little research or attention has been given to the implications of unstable population dynamics on evolution in age-structured populations, although there may be many populations in which these conditions are met.

WHY CONDUCT LABORATORY EXPERIMENTS?

For most other branches of biology, this question would seem naïve at best. To control variables except those of interest, replicate experiments under well-defined conditions seem obvious and necessary. Yet in ecology there is a long tradition and interest in collecting observations and conducting experiments in uncontrolled, or semi-controlled, natural environments (Carpenter, 1996). The most compelling argument for this tradition is that ecology is ultimately interested in the factors that affect the abundance and distribution of plants and animals in their natural environment. Consequently, many believe that observations in the laboratory will not add to our understanding of nature (Peters, 1991).

This type of argument is not confined entirely to ecology. Much research in aging, for instance, has focused on the aging of individual cells. However, the utility of this information for inferring the mechanisms that affect the aging of whole organisms is debatable. One reason why the behavior of aging cells may not inform us about whole-organism senescence is that the process of aging is largely determined by natural selection and may be, therefore, expected to have heterogeneous causes among both organ systems and species (Rose, 1991).

We feel, however, that the argument against laboratory experiments in ecology is far less compelling than that just outlined for the irrelevance of studying cellular aging to understanding organismal senescence. There is a tendency in many discussions in ecology to speak of "nature" as if it were a single, well-defined set of conditions. In reality, of course, the natural environment is heterogeneous over time and space (fig. 1.2). Thus, the "natural" environment of a species in one year or season may be quite different from the "natural" environment in the next year or season. In that sense, among the infinite hierarchy of environments, the laboratory may be no more or less special than any other.

Certainly our ability to develop and test theories in ecology is not at an advanced stage. We are not yet at the point where our theories can incorporate and deal with the myriad of variables that are constantly changing in nature (Drake et al., 1996). The impatient ecologist will often suggest that perhaps the many details do not matter, and only competition, or only predation matters; thus, they are willing to test their theory with data from natural populations. This approach typically requires choosing data sets carefully and developing an ad hoc rationale for dismissing apparent contradictions to theory.

In our opinion, the laboratory is one of the best places to rigorously test ecological theory. Many of these theories are bound to fail even in the simplified environment of a laboratory. The ability to improve and reconstruct theory, however, will require that we understand why our prior theories have failed. In the laboratory the ability to determine the causes of theory failure will almost always be more straightforward and easy to diagnose than in the field.

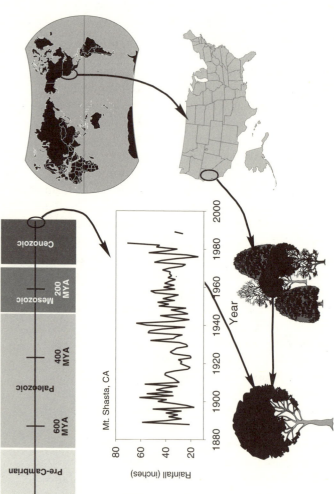

FIG. 1.2. Temporal and spatial scales of ecological studies. This figure emphasizes the time and spatial scale that ecological studies represent by showing the ever-narrowing range of a particular field site. A tree and its insect flora in Mt. Shasta, California, is just one of several trees in the immediate forest, which may be one of many forests in the world. Likewise, the observable events in one year may or may not be characteristic of the past 100 years, which may be quite different from those of earlier epochs.

LABORATORY STUDIES OF POPULATION BIOLOGY

The use of laboratory studies in ecology and evolutionary biology has a long history. Laboratory experiments permit certain aspects of the environment to be controlled and thus remove confounding factors that exist in natural populations. However, there are a variety of issues that need to be considered when designing laboratory experiments; we review these here. Although much of our focus is on ecological problems, many of the issues we address have been considered previously by Rose et al. (1996) in their discussion of laboratory studies of evolution.

Starting Populations

Many problems in ecology and evolution are concerned with populations that are normally outbreeding and genetically variable. This means that the populations brought into the laboratory should also be genetically variable. Thus, the original samples should be as large as practically possible. Laboratory experiments are sometimes started with isofemale lines. Isofemale lines are initiated by placing single, inseminated females in their own culture. Usually the progeny from these females then mate with each other for one or more generations. Even if the isofemale lines are later pooled together, the inbreeding and subsequent crossing lead to high levels of linkage disequilibrium that are unlikely to characterize populations in nature.

Some studies have used laboratory mutants or captive wild-type stocks, especially studies using *Drosophila* and *Tribolium*. Visible morphological mutations often have pleiotropic effects on many fitness traits that may affect population dynamics (Prout, 1971a; Bundgaard and Christiansen, 1972). Laboratory mutant and captive wild stocks (such as *Oregon-R* in *Drosophila* or *Canton-S* in *Drosophila*), often have irregular or unknown maintenance histories that can include frequent episodes of population bottlenecks or chronic maintenance at low population size. Consequently, most laboratory stocks, whether mutant or wild-type lines, make poor starting material for experiments in ecology and evolution. The exception to this is laboratory stocks that have been maintained consistently at large population sizes under a well-defined and carefully controlled maintenance regime.

This problem can be illustrated with the following hypothetical example. Suppose a human population was inbred and used for experimental research. In this particular population, the inbreeding gave rise to a group homozygous for the sickle-cell anemia allele. Suppose this group is now compared to a second inbred population homozygous for the normal hemoglobin allele (e.g., wild type). Certainly a lot can be learned about the how the hemoglobin protein works by extracting hemoglobin from each of these two groups. However, if you were unaware of the nature of the genetic differences between these two groups, you might be tempted to argue that the wild-type group contained alleles that increased longevity and thus were the key to reversing aging (since those with sickle cell anemia do not live as long as normal individuals). You could also incorrectly conclude that the alleles in the wild-type population were the key to understanding stress resistance, since those individuals could outperform the sickle-cell population in a wide variety of activities that require aerobic endurance. But in general, the comparison of these two inbred populations tells us nothing about the genetic or physiological factors that affect longevity or stress resistance in normal populations. Rather, they reveal the effects of a rare genetic disorder that has been made common in one population through inbreeding.

As a real example, consider inbred genotypes of *D. melanogaster* derived by the technique of chromosome extraction (Mueller and Ayala, 1981d). These inbred genotypes show strong positive correlations in rates of populations growth: Those genotypes that grow quickly at low density also grow quickly at high density, and those that grow slowly at one density tend to grow slowly at all densities (Mueller and Ayala, 1981d). What can we infer from these observations about the evolution of population growth rates, or about high fitness genotypes, in natural populations? The answer turns out to be very little. In fact, when genetically variable *D. melanogaster* populations are maintained for many generations at either high or low population density, those genotypes that rise in frequency and become predominant exhibit trade-offs, in stark contrast to the result obtained from the inbred lines: Those genotypes that do best at low density grow more slowly at high density, and vice versa (Mueller and Ayala, 1981a; Mueller et al., 1991). Thus, the properties of inbred genotypes bear no resemblance to the high fitness genotypes in the outbred populations that are ultimately favored by natural selection operating at different densities.

Lab Adaptation

Studies in which the behavior of populations is tracked over many generations must also consider the possible effects of evolutionary change in these populations. In many ecological studies, such evolution may be undesirable because it may change important properties of the population that the experimenter wishes to keep constant. However, wild populations brought into the laboratory inevitably undergo evolutionary change as they adapt to food, temperature, crowding, and other aspects of the laboratory environment that differ from the natural environment of that population in the field. For instance, natural populations of *Drosophila* brought into the laboratory show a continual increase in adult population size as they adapt to the laboratory (Buzatti-Traverso, 1955; Ayala, 1965b, 1968). Consequently, laboratory studies aimed at testing evolutionary or ecological theories may be thwarted if the experimental and control populations are still adapting to features of the laboratory environment. For starting material in laboratory studies in ecology and evolution, therefore, it is most desirable to use large, outbred populations that have had time (12 generations or more) to adapt to the laboratory environment.

Replicate Populations

Most experimental research in population dynamics and evolution uses whole populations as the units of observations. Consequently, the power of any analysis is a function of the number of replicate populations. Experiments with one experimental and one control population have no power. Often the maximum number of populations is determined by practical factors such as time and cost of maintenance. However, with five replicate controls and five experimental populations, one can meet the minimum sample size requirements for several non-parametric tests (e.g., Wilcoxon's signed-ranks test, Sokal and Rohlf, 1981, p. 448).

In ecological studies, replicates serve the traditional role of ensuring that observed differences between experimental and control populations are a consequence of the experimental conditions and not random, uncontrolled factors. In evolutionary studies, however, the importance of replicates takes on a whole new dimension. Genetic differences may always arise between two populations due to random genetic drift. However, most laboratory studies are interested in genetic differentiation due to natural selection. Thus, the observation of genetic differences between one control population and one experimental population does not help us distinguish between the relative importance of selection and drift as causative agents. However, it is unlikely that five or more replicate populations will experience the same sequence of random events. Thus, in evolutionary experiments, the key to distinguishing between drift (a stochastic force) and selection (a deterministic force) is the observation of consistent differentiation among multiple independent populations.

A related problem is the size of the replicate populations. If the primary interest of the study is to investigate the outcome of natural selection, then the replicate populations ought to be as large as possible. There are two reasons for this recommendation: (1) In large populations, selection can act effectively on alleles that have small effects on fitness. Consider a locus with two alleles and hence three genotypes: A_1A_1, A_1A_2, and A_2A_2. Let the fitness be additive and equal to $1 + s, 1 + 1/2s$, and 1 respectively. If the initial frequency of the favored A_1 allele is p, then fixation is virtually assured if $N_esp > 5$, where N_e is the effective population size (Ewens, 1979, p. 147). As an example, in a population with $N_e = 1000$, in which a favored allele exists as only a single copy, s would have to be 10 to be assured of fixation. Thus, in most laboratory experiments we cannot be certain that very rare favorable mutants will be fixed. However, an allele that is at a frequency of 10 percent is virtually guaranteed of fixation if the favored homozygote has a 5 percent or greater advantage over the alternative homozygote. If the population size had been 100 rather than 1000 the favored homozygote would have required a 50 percent fitness advantage rather than 5 percent. As N_e, s, and p become smaller, the chance of a favored allele becoming fixed decreases, and conversely the chances increase that the disfavored allele will be fixed. When $N_es < 0.1$, the chance of fixation is very close to that of a neutral allele.

(2) Small populations increase the chance that deleterious alleles will be fixed and thus potentially obscure the effects on fitness of beneficial alleles at other loci. This is especially likely to be the case when life history traits are examined, since there appears to be abundant genetic variation that has deleterious pleiotropic effects on survival and fertility (Lewontin, 1974). For neutral alleles that are fixed by drift it takes on average four N_e generations for fixation. Alleles that directly affect fitness negatively, take longer on average to be fixed, although their ill effects are apparent well before they are fixed in a population.

These concerns extend to ecological studies as well. Drift and inbreeding may affect life history traits that ultimately affect population dynamics in a significant fashion. Population stability, for instance, is often dependent on female fecundity. However, inbreeding may significantly reduce female fecundity and thus the dynamics of a population about equilibrium (see chapter 6 for a detailed discussion).

Measuring Genetic Differences

Many laboratory studies with an evolutionary component ultimately need to determine if there are genetic differences among populations. Often the interest is not in the particular frequencies of alleles in each population but in the unknown alleles and loci that affect quantitative traits. These traits are often affected by the environment and sometimes by the maternal environment. In *Drosophila*, for example, the level of larval crowding affects the ultimate size of the adult, with small adults emerging from crowded cultures. However, phenotypes, such as longevity (Miller and Thomas, 1958) and fecundity (Chiang and Hodson, 1950) are affected by adult size. Smaller adults tend to live longer, and smaller females lay fewer eggs. The maternal environment may also be important. Egg-to-adult viability in *Drosophila* is reduced as parental age increases (Rose, 1984).

These effects can be removed by rearing test organisms for two generations in a common environment. One would sample adults directly from the control and the experimental populations and let them produce offspring under common conditions. The progeny that emerge from this generation will all have experienced a common environment but may differ due to differences in their parents' ages or nutritional states. Thus, one more generation is needed to obtain juveniles or adults that can be assayed for phenotypes of interest. If there are significant differences between experimental and control populations in the second-generation individuals, these can be attributed to underlying genetic differences between the two populations.

EVALUATING MODELS IN POPULATION BIOLOGY

The analysis of population stability inevitably requires some characterization of the study organism's population dynamics. This characterization will often be in the form of a mathematical model. We use the word *model* to generally mean an abstraction and typically a simplification of a biological process. From this definition it is clear that a model need not be mathematical but could be a verbal description of the abstraction. Of course the virtue of mathematical models is that their implications may be studied by the formal and generally understood techniques of mathematical analysis.

Some mathematical models may represent important biological theories. When considering population dynamic models, for instance, we may construct them by careful consideration of the life history of a particular organism and the various ways these life histories are affected by biological attributes such as density or age. A comparison of the predictions of these sorts of models to empirical observations is then, to some extent, a test of our biological understanding of life history.

Models can also be constructed from simple statistical techniques. Thus, the dynamics of a population may be modeled by a high-order polynomial whose coefficients have no biological meaning but have been estimated from a set of observed population trajectories. In either case, the utility of a model may ultimately be assessed by comparing its predictions with a set of observations. The manner in which this is done varies greatly. With reference to population dynamics, comparing goodness of fit to some existing data set may help assess different models. Alternatively, the ability of the population dynamics model to predict new observations may be used as a criterion for model selection. In some cases the model may predict unusual behavior in altered environments, and these predictions can be tested experimentally.

Royama (1971) has reviewed some of the major factors that may lead to differences between a model's predictions and empirical observations. Royama makes the obvious but sometimes unappreciated point that such differences do not always mean that the model is wrong. A model can be viewed as consisting of its components and its structure. For a population-dynamic model, the components might be pre-adult survival, adult survival from one age class to the next, adult fertility, etc. To define the structure of the model, using this same example, we would need to specify how the different components of the model interact. If survival shows density dependence, does it change in a linear fashion with density or in some nonlinear fashion? Clearly, there may be differences between the predictions of a model and observations from experiments or field populations, because *the components or the structure of the model may be wrong or insufficient*. However, differences between the model predictions and observations may also arise because *the conditions under which the observations were collected violate specific and important assumptions of the model*. Thus, a natural population may never shown a sustained and constant equilibrium population size because the level of essential resources varies over time and is not constant as assumed by a simple model of population dynamics.

The remedy to take in each of these cases is quite different. If we can reasonably conclude that the model is wrong, then we need to adjust it in a way that is suggested by our experimental findings. However, if the original observations are suspect, then we need to find or design an experimental system that can adequately test the model. Unfortunately when observations have been made in natural populations, it is often too easy to invoke the uncontrolled aspects of the environment as the culprit for a model's failure. Well-designed laboratory experiments should permit us to reject models only when their predictions are discordant with observations. This is ultimately the power of strong inference (Platt, 1964).

GENERAL VERSUS SPECIFIC MODELS

An important component of all scientific research is the transition from theoretical predictions to experimental tests. The theory of population genetics and ecology often assumes discrete generations and populations without stage or age structure. Problems arise when these theorics are tested with organisms that depart from these assumed life histories. Organisms such as *Drosophila*, for instance, can be made to reproduce on a discrete schedule, and adult age classes can be eliminated, but the prereproductive stages of *Drosophila* can never be removed. Attempts to estimate fitness coefficients from simple population genetic models with organisms such as *Drosophila* can be thwarted if selection acts on the different components of the life cycle (Prout, 1965, 1971a, 1971b). This coupled with the necessity to assay adults rather than eggs means that the most general models of selection are inappropriate for providing a framework for observations in the simplest of *Drosophila* populations.

Prout has also recognized that similar problems occur in simple models of population dynamics (Prout, 1980; Prout and McChesney, 1985; Prout, 1986). For instance, the simplified life cycle of *Drosophila* in the laboratory always has three different census stages: larvae, pupae, and adults. A model keeping track of population size might refer to any one of these life stages. If selection acts in a density-independent fashion, it is possible for evolution to increase, decrease, or have no effect on equilibrium numbers of particular census stages. Any general claim that selection will always maximize population size is not true.

Prout has also noted that fertility in some organisms depends on pre-adult density. Crowding during these stages often has lasting effects on adult size that in turn affect fertility. This biological phenomenon posses difficult problems for estimating the underlying population dynamics

31

from data on adult numbers only. Although this problem is not insurmountable (we discuss some solutions to it in chapter 2), it must be considered in the development of model experimental systems.

These issues raise the general question of the most appropriate type of model to use when developing theory in life-history evolution. Christiansen (1984) makes a distinction between phenomenological and explanatory models. The phenomenological models are simple and attempt to summarize the totality of density dependence or other factors with a single simple function (e.g., the logistic). For these reasons the models are thought to have greater generality (Levins, 1968). Whereas explanatory models explicitly take into account specific components of the life cycle of some organism or group of organisms and try to model the response of these life history components to density, parasites, etc. These latter models have less generality since the life-history details included in these models may vary from one taxonomic group to another. Christiansen argues that this is the most appropriate way to develop theory for the study of life-history evolution in variable environments. Certainly, if theory is being used to make specific predictions about the evolution of a particular population, one cannot use a model that ignores crucial life-history details.

Despite the simplicity of the laboratory environment, the design of good experiments with model systems is a multifaceted affair and must be planned carefully. Once the careful planning is done, however, there are great benefits to be derived from experiments with model systems. We use the remainder of this book to develop some of the knowledge about population stability that has been learned from experimental work on model systems in the laboratory.

CHAPTER TWO

Theory of Population Stability

Many of the problems associated with population dynamics were originally suggested by the analysis of simple models. These models may often be unrealistically simple; however, they are useful starting points in the exploration of population dynamics, and they have substantial heuristic value. In this chapter we review the classical mathematical techniques for determining the stability characteristics of these simple models. By devoting attention to specific methods, the mathematical meaning of stability should become apparent. In addition, these methods will aid in our discussion of the various techniques that have been suggested to determine population stability empirically, since many of these techniques mimic the mathematical analysis of stability. In this chapter we focus on the stability concepts for deterministic models. In the next chapter, the concepts of stochastic stability are reviewed.

The application and use of simple models requires careful evaluation of important life-history features of specific organisms. We review the consequences of age structure and interactions among different life stages on the ability to infer population stability. These issues are important since it is often not possible to collect all relevant information from laboratory or natural populations. For example, one often has information on total numbers but not the number of individuals in each age class or sex. We need to know if we can make accurate inferences concerning population stability with incomplete data and, if not, how seriously the lack of various specific details affects our ability to draw inferences about population stability.

33

Populations typically harbor genetic variation for life-history traits that can directly affect population stability (Mueller and Ayala, 1981c). Consequently, population stability may evolve in concert with these life-history traits, possibly as a by-product of the evolution of these traits. Accordingly, the conditions that foster the evolution of population stability may help us interpret observed patterns in natural populations (Turchin and Taylor, 1992).

FIRST-ORDER NONLINEAR DIFFERENCE
AND DIFFERENTIAL EQUATIONS

A simple model is often one that has few parameters. For models of population growth, this usually means that the size of the population is assumed to depend on only one immediate past population size. If an organism reproduces continuously and all members of the population are considered to be equivalent, then the most appropriate description of population dynamics is through the use of differential equations in continuous time. If we let N be the total number of individuals in the population, then the rate of change of this number, dN/dt, depends on the current population size according to some function $f(N)$. When $f(N)$ is a nonlinear function, the resulting model is a nonlinear differential equation. Nonlinear differential equation models clearly presume that the effects of density on reproduction and survival are instantaneous, which may, in fact, seldom be the case for many populations.

An alternative modeling approach is to assume that reproduction in the population is synchronized but is preceded by a period of development or, at least, an absence of reproduction. Following reproduction, the adults may all die, leaving only the progeny to form the next generation. Alternatively, we may assume that some fraction of the adults survive to the next generation. In these discrete-time models, however, as with the continuous-time models, all members of the population are considered equivalent. Thus, adults who survive must have the same capacity for reproduction as do the newborn progeny (i.e., there is no adult age structure). Because time can be viewed meaningfully as changing in discrete steps, the general form of these models is of a difference equation wherein population size at time t depends on that at time $t - 1$ ($N_t = g(N_{t-1})$). The function $g(N_{t-1})$ can be decomposed into two parts, $N_{t-1} \times$ (the per-capita growth rate, $\lambda(N_{t-1})$). The per-capita growth rate is typically assumed to decline with increasing population size due to biological factors such as density-dependent survival and fertility (Begon et al., 1990, pp. 206–209). The precise form of the decline differs among models and, in the simplest case, can be assumed to be linear (although the function $g(N_{t-1})$ is still nonlinear). We can illustrate this type of model with three different formulations of density dependence of per-capita growth rates (Prout, 1980):

$$\lambda(N_t) = a_1 + a_2 N_t \text{ (linear)} \tag{2.1}$$

$$\lambda(N_t) = \frac{a_1}{1 + a_2 N_t} \text{ (hyperbolic)} \tag{2.2}$$

$$\lambda(N_t) = a_1 \exp[a_2 N_t] \text{ (exponential)} \tag{2.3}$$

The linear model gives rise to a model of population growth called the linear logistic or the quadratic map. Typically, the linear logistic equation is presented with two parameters: r(which equals $a_1 - 1$), the intrinsic rate of growth; and K (which equals $(1 - a_1)/a_2$), the carrying capacity. This model may also be derived by considering the reproduction and dispersal of single individuals (Lomnicki, 1988). The exponential formulation yields a model of population growth variously called the exponential logistic or Ricker map. Parameters of the three models have been estimated from observed population sizes in a single population of *Drosophila melanogaster* (fig. 2.1). The predictions from all three models give reasonable descriptions of the observed population sizes. One point we wish to stress is that the simple observation of concordance between observed and

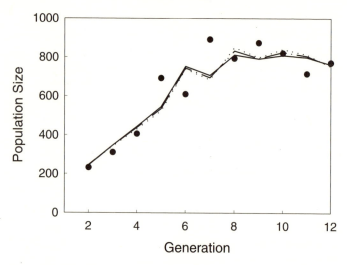

FIG. 2.1. Number of adults (filled circles) in the HL_1 laboratory population of *Drosophila melanogaster* studied by Mueller and Huynh (1994). The lines show the predicted population sizes from models 2.1 through 2.3 based on maximum-likelihood estimates (Dennis et al., 1995; see box A). The solid line is the predicted population size from the logistic equation (2.1); the dotted line is the predicted population size from the hyperbolic equation (2.2); and the dot and dashed line is the predicted population size from the exponential model (2.3).

predicted population sizes by itself provides only weak support for a model. To gather strong support for a theory requires one or more of the following types of observations: (1) obtaining estimates of the model parameters independently of the population dynamics and then producing accurate predictions; (2) correctly predicting qualitatively different dynamics that would be expected under certain conditions and which the model correctly predicts (e.g., among models 2.1 to 2.3, the logistic and exponential models can produce cycles and chaos, whereas the hyperbolic model cannot); or (3) using the model to make predictions about other aspects of the population (e.g., numbers of other life stages or average adult size), which can then be used to verify the model independently.

BOX A
MAXIMUM LIKELIHOOD:

Suppose we have a sample of n, random variables, x_1, $x_2, \ldots, x_n$. These may be either discrete or continuously varying random variables. The probability density function is assumed to depend on the value of the random variable and a parameter θ, and is represented by $f(x_i|\theta)$. We then define the likelihood function for a particular sample and values of θ as follows:

$$L(\theta) = \prod_{i=1}^{n} f(x_i|\theta)$$

If x_i is a discrete random variable, the likelihood function is equal to the probability of drawing the observed sample. The maximum-likelihood estimate of θ is designated $\hat{\theta}$ and is chosen to maximize the likelihood function. In the simplest cases, we can use elementary calculus to find the value of θ that satisfies the following equation:

$$\frac{\partial L(\theta)}{\partial \theta} = 0$$

Continuous-time versions of models 2.1 to 2.3 can be produced by deriving expressions for $N_{t+1} - N_t$. A differential equation can be obtained by letting this time difference go to zero. In that case the three new models can be written using the notation in equations 2.1 to 2.3:

$$\frac{dN}{dt} = N(a_1 - 1 + a_2N) \quad \text{(linear, often referred} \quad (2.4)$$

to as the continuous-time logistic model)

$$\frac{dN}{dt} = N\left(\frac{a_1 - 1 - a_2N}{1 + a_2N}\right) \quad \text{(hyperbolic)} \quad (2.5)$$

$$\frac{dN}{dt} = N\left[a_1 \exp(a_2N) - 1\right] \quad \text{(exponential)} \quad (2.6)$$

37

Box A cont.

Often it is easier to find the maximum if we first take the log of the likelihood function. Some density functions may have k random variables rather than one variable that is subject to constraints, as in $\sum_{i=1}^{k} \theta_i = 1$. These problems require a more sophisticated method of finding the maximum, called *Lagrange multipliers* (Intriligator, 1971, p. 28). If the derivative of the likelihood function cannot be solved analytically, numerical methods must be used to find the maximum (Beveridge and Schechter, 1970).

Dennis and Taper (1994) discuss applications of maximum-likelihood techniques to population data. Suppose we have a time series of $m + 1$ population sizes, $N_0, N_1, \ldots, N_m$. We believe the dynamics of this population to be governed by a simple first-order difference equation, $N_t = N_{t-1} g(N_{t-1})$. If we let $x_t = \ln(N_t)$, then random noise can be introduced into this equation as

$$x_t = x_{t-1} + \ln\big[g(N_{t-1})\big] + z_{t-1},$$

The right-hand side of each equation (2.4–2.6) has N multiplied by a term in parentheses. This term is no longer a per-capita growth rate but is the instantaneous increase in population size due to the difference between births and deaths. To find equilibrium points of the discrete-time models, we determine the population size at which the per-capita growth rates are equal to one (i.e., the density at which each individual in the population can just replace itself). To find the equilibrium of the continuous-time model, one must determine the population size at which the instantaneous increase is exactly zero, meaning that births just balance deaths. An important property of these models, both discrete time and continuous time, is whether the equilibria just described are stable. We next review the standard mathematical techniques for answering this question.

BOX A cont.

where z_t is assumed to be normally distributed with mean 0 and variance σ^2. Next, compute the density function of x_t conditional on the previous log-population size, x_{t-1} as follows:

$$f(x_t|x_{t-1}) = \frac{1}{\sqrt{2\pi\sigma^2}} \exp\left\{ \frac{\left(x_t - \left(x_{t-1} + \ln[g(N_{t-1})]\right)\right)^2}{2\sigma^2} \right\}$$

The likelihood function is then defined as follows:

$$L(\theta) = \prod_{i=1}^{m} f(x_t|x_{t-1})$$

$$= \frac{1}{(2\pi\sigma^2)^{m/2}} \exp\left\{ -\frac{1}{2\sigma^2} \sum_{i=1}^{m} \left(x_i - x_{i-1} - \ln g[N_{i-1}]\right)^2 \right\}$$

The derivative of this function must be taken with respect to σ^2 and the parameters of the function $g(N_t)$. The resulting equations are set to zero and their solutions found.◆

STABILITY OF FIRST-ORDER NONLINEAR DIFFERENCE AND DIFFERENTIAL EQUATIONS

In this section we focus mainly on the analysis of local stability, implying that the statements concerning the behavior of the dynamic system will only hold in a small region close to an equilibrium point. In contrast, a globally stable equilibrium is approached from all feasible population sizes. Conceptually, analyzing local stability involves determining the dynamics of the system in a region close to the equilibrium point of interest. The word *close* for our discussion means that the range of population sizes examined is sufficiently narrow that we can approximate the nonlinear functions describing the dynamics with linear functions. The mathematical technique used to produce this linear approximation is called a Taylor series expansion. An outline of Taylor's theorem is given in box B. The subsequent discussion of the stability of

BOX B

TAYLOR SERIES:

Taylor's theorem provides a convenient means of estimating certain types of complicated functions. If the function can be differentiated, then in principle the function can be approximated to any desired degree of accuracy. If we consider only functions of a single variable, x, then we also need to choose a single value of x, x^*, that will be close to the values of x we wish to use in our function. The estimates provided by Taylor's theorem are most accurate when x is close to x^*. The level of accuracy depends on both the form of the function and how many terms in the Taylor series are used. If we let the function be $f(x)$ and $f^{(n)}(x^*)$ be the nth derivative of $f(x)$ evaluated at the point x^*, then Taylor's theorem says

$$f(x) = f(x^*) + \frac{(x - x^*)}{1!} f^{(1)}(x^*) \qquad (2.7)$$

$$+ \frac{(x - x^*)^2}{2!} f^{(2)}(x^*) + \cdots + \frac{(x - x^*)^n}{n!} f^{(n)}(x^*)$$

$$+ \frac{(x - x^*)^{n+1}}{(n + 1)!} f^{(n+1)}(\xi_x),$$

growth models can be found in many elementary texts, and a particularly nice example for population growth models is given in Roughgarden (1979).

In our discussion of local stability analysis, we first consider continuous-time growth models. An equilibrium for these models must satisfy the condition that $dN/dt = f(\hat{N}) = 0$. The problem we must solve is to describe the behavior of a small perturbation, ε, to this equilibrium, $N = \hat{N} + \varepsilon$. Does the perturbation die off to zero and return the system to the equilibrium population size $\hat{N}$, or does it increase in magnitude and move the population away from the equilibrium? To study this we approximate the effect of the perturbation on population size, $f(\hat{N} + \varepsilon)$, with a Taylor series expansion

---

BOX B cont.

where ξ_x is some point on the interval where $f(x)$ is defined. The last term in equation 2.7 is called the remainder, and its exact value is typically unknown. The Taylor series approximation (or expansion) of the function $f(x)$ is all terms on the right-hand side of equation 2.7 except the remainder. As an example, consider the exponential function e^x. The value of the nth derivative is always e^x, for all values of n. If we center the Taylor series approximation around the point, $x^* = 0$, then we have the following approximation:

$$e^x \cong 1 + x + \frac{x^2}{2!} + \frac{x^3}{3!} + \cdots + \frac{x^n}{n!} \qquad (2.8)$$

If we use just the first two terms in equation 2.8, our approximation of $e^{0.1}$ is 1.1 while the exact value is 1.105. However, as we try to get estimates with values of x further from zero, the accuracy of the prediction decreases. Thus, the approximation to $e^{1.1}$, using just two terms of the Taylor series, is 2.1 while the exact value is 3.00.◆

---

about the point, $\hat{N}$, and use just the first two terms in the series:

$$\frac{dN}{dt} = \frac{d(\hat{N} + \varepsilon)}{dt} \qquad (2.9)$$

$$= \frac{d\varepsilon}{dt} \cong f(\hat{N}) + f^{(1)}(\hat{N})\varepsilon = f^{(1)}(\hat{N})\varepsilon$$

Equation 2.9 can then be integrated to find the time-dependent behavior of ε:

$$\varepsilon(t) \cong \varepsilon(0)e^{f^{(1)}(\hat{N})t} \qquad (2.10)$$

Consequently, if the first derivative, $f^{(1)}(\hat{N})$, is less than zero, then $\varepsilon(t) \to 0$ as $t \to \infty$ (read the symbol "$\to$" as "goes to"). That is, the population returns to the equilibrium, $\hat{N}$. If $f^{(1)}(\hat{N})$ is greater than zero, then $\varepsilon(t) \to \infty$ as $t \to \infty$. The population size departs from the equilibrium at an exponential rate of increase, at least initially. As an example, for the linear or logistic model (equation 2.4), $f^{(1)}(\hat{N}) = 1 - a_1$ or r. Thus, local stability of the equilibrium, $\hat{N}$, is ensured if $a_1 > 1$, or $r > 0$. Because r is the maximal per-capita instantaneous rate of increase, under ideal conditions, it is positive unless the population is declining in numbers with time and is inviable in the long run. Thus, the continuous-time logistic model predicts a stable equilibrium for any increasing population.

In discrete-time models, an equilibrium must satisfy the condition $g(N_t) = \hat{N}$. As before, we need to examine a perturbation to the equilibrium, ε_t. We study the time-dependent behavior of this perturbation by noting the following equation:

$$\hat{N} + \varepsilon_{t+1} = g(\hat{N} + \varepsilon_t) \cong g(\hat{N}) + g^{(1)}(\hat{N})\varepsilon_t$$
$$= \hat{N} + g^{(1)}(\hat{N})\varepsilon_t.$$

If we subtract $\hat{N}$ from both sides, we get the following solution:

$$\varepsilon_{t+1} \cong g^{(1)}(\hat{N})\varepsilon_t = \left[g^{(1)}(\hat{N})\right]^{t+1}\varepsilon_0 \qquad (2.11)$$

Thus, equation 2.11 predicts that stability will be ensured if $|g^{(1)}(\hat{N})| < 1$. If $|g^{(1)}(\hat{N})| = 1$. Further analysis is required, and if $|g^{(1)}(\hat{N})| > 1$, the equilibrium is unstable. Sometimes the quantity, $g^{(1)}(\hat{N})$, is referred to as the leading or stability-determining eigenvalue.

Using this approach, we have determined the equilibrium population size and stability-determining eigenvalue for that equilibrium for each of the discrete time models 2.1–2.3 (table 2.1). It is worth noting that, for biologically reasonable values of the parameters a_1 and a_2, the linear logistic and the exponential model can produce eigenvalues of absolute value greater than one, whereas the hyperbolic model cannot. Careful examination of equation 2.2 shows that the parameter a_1 will equal the per-capita growth rate of the population when N is very small. For that reason it must at least be positive. In fact, a_1 must also be greater than one

TABLE 2.1. Equilibrium population size, $\hat{N}$, and the stability-determining eigenvalue, λ, for the discrete-time models 2.1 to 2.3.

Model	$\hat{N}$	λ
Linear logistic	$\dfrac{1 - a_1}{a_2}$	$2 - a_1$
Hyperbolic	$\dfrac{a_1 - 1}{a_2}$	$\dfrac{1}{a_1}$
Exponential	$-\dfrac{1}{a_2}\ln(a_1)$	$1 - \ln(a_1)$

or else the equilibrium at $N = 0$ is stable (i.e., the population goes extinct). With $a_1 > 1$, the stability-determining eigenvalue for the hyperbolic model is always less than one. Hence the hyperbolic model predicts that all feasible equilibria with $\hat{N} > 0$ will be stable.

We illustrate models 2.1 to 2.3 with data from a population of *D. melanogaster* (fig. 2.2) exhibiting very different growth characteristics from those of the population illustrated in figure 2.1. We discuss the causes of these differences in more detail in chapter 6. What is clear from the figures themselves is that the population in figure 2.2 is fluctuating more violently than the population in figure 2.1 and, over the nine generations of observations shown, has not settled down to what might be considered an equilibrium population size.

For the populations illustrated in figure 2.1 (HL$_1$) and figure 2.2 (LH$_1$), we present maximum likelihood estimates for the parameters of models 2.1–2.3 (table 2.2). Of interest is the value of the stability determining eigenvalue, λ, predicted by each of these models. For all models, $|\lambda| < 1$ for population HL$_1$ (fig. 2.1), suggesting that a stable equilibrium exists. However, for the population LH$_1$ (fig. 2.2), the linear and exponential models predict that the equilibrium population size, $\hat{N}$, is unstable.

This brief analysis of data illustrates an important point concerning the use of models. Based on the results shown in

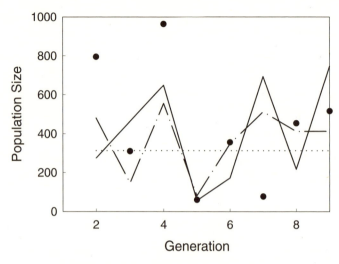

FIG. 2.2. Number of adults (filled circles) in the LH$_1$ population of *Drosophila melanogaster* studied by Mueller and Huynh (1994). The lines show the predicted population sizes from models 2.1 through 2.3 based on maximum-likelihood estimates (Dennis et al., 1995). The solid line is the predicted population size from the logistic equation (2.1); the dotted line is the predicted population size from the hyperbolic equation (2.2); and the dot and dashed line is the predicted population size from the exponential model (2.3).

figure 2.1, one is tempted to conclude that the three models (2.1–2.3) all do an adequate job describing the dynamics of these *Drosophila* populations. However, as discussed in chapter 1, simple agreement between observations and predictions, especially when the predictions have relied on the observed data to some extent, is not strong support for a particular model. Consequently, there must be a vigorous search for methods independent of simple model fitting for validating population dynamic models. We see through the analysis of the LH$_1$ population data that the hyperbolic model is not capable of providing an adequate description of its dynamics. It seems sound to conclude, therefore, that the hyperbolic model is not a generally useful model for describing *Drosophila* population dynamics.

TABLE 2.2. Parameter estimates for three models and two populations of *Drosophila melanogaster*, HL_1 and LH_1, shown in figures 2.1 and 2.2, respectively.

Model	Population	Model Parameter Estimates			
		a_1	a_2	$\hat{N}$	λ
Linear	HL_1	1.701	-0.000885	792	0.299
	LH_1	3.06	-0.0031165	661	-1.06
Hyperbolic	HL_1	1.83	0.00103	805	0.546
	LH_1	2.05×10^8	6.55×10^5	313	4.88×10^{-9}
Exponential	HL_1	1.76	-0.000710	796	0.435
	LH_1	7.69	-0.0047	434	-1.04

POPULATION CYCLES AND CHAOS

Cycles

A question that arises from the preceding discussion of stability is what happens in the case of populations controlled by models 2.1 and 2.3 when their equilibrium points are unstable. The moment the equilibrium points listed in table 2.2 become unstable, two new stable equilibria appear, and the population begins to cycle between them. This phenomenon is known as period doubling, or bifurcation (May, 1974; May and Oster, 1976). Since the equilibrium point is now unstable, it acts as a repellor, meaning that points close to it move away. Since the instability is due to the eigenvalue becoming less than −1, points near the former equilibrium will oscillate above and below the previous equilibrium as they move away from it. As a result of this behavior this process is sometimes called flip bifurcation (Hilborn, 1994).

As the value of the parameter a_1 in models 2.1 and 2.3 continues to increase, a threshold value is reached at which the two-point cycle itself becomes unstable, and each of these equilibria further bifurcates to produce a stable four-point cycle. These period doublings continue until there are an infinite number of period doublings, and the population exhibits a form of dynamics called chaos. For the linear model, the transition to chaotic dynamics occurs when $a_1 = 3.57$. We might qualitatively describe chaos as unstable, aperiodic behavior (Kellert, 1993). Period doubling is just one of several pathways to chaos. We review these in more detail in the next section. To gain some additional insights into the properties of these new equilibria, let us first consider the stability of the pair of equilibria that appear when a_1 just exceeds 3.0 in the linear model.

If we iterate models 2.1 and 2.3 for the parameter estimates obtained for the LH$_1$ population (table 2.2), each model appears to settle into a two-point cycle (fig. 2.3). If we label these new equilibrium points $\hat{N}_1$ and $\hat{N}_2$, then

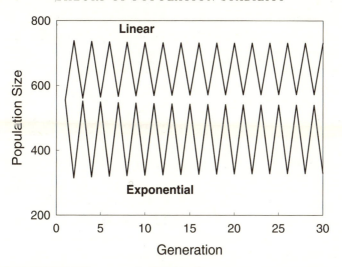

FIG. 2.3. Variation in population size for the linear (2.1) and exponential (2.3) models. The parameter values used were those estimated for the LH$_1$ population in table 2.2.

from model 2.1, the linear model must satisfy the following equations:

$$\hat{N}_1 = \hat{N}_2(a_1 + a_2\hat{N}_2) \qquad (2.12a)$$

$$\hat{N}_2 = \hat{N}_1(a_1 + a_2\hat{N}_1) \qquad (2.12b)$$

Substituting 2.12b into 2.12a yields a cubic equation in $\hat{N}_1$, which has three solutions. One of these solutions is the previous equilibrium, $(1 - a_1)/a_2$, which is now unstable. The other two equilibria are the points seen in figure 2.3, which in general must be determined numerically. For the LH$_1$ population, the equilibria are $\hat{N}_1 = 572$ and $\hat{N}_2 = 730$ for the linear model, and $\hat{N}_1 = 330$ and $\hat{N}_2 = 538$ for the exponential model.

We next turn to the question of the stability of this two-point cycle. Formally the mathematical analysis of stability for the two-point cycle is done as we have outlined for a single-point equilibrium. The core determinant is, when

47

the system is perturbed slightly from this two-point cycle, whether it returns to the cycle or moves away (see May and Oster, 1976, for a more detailed description). In the case of a two-point cycle, the stability-determining eigenvalue, λ_2, is given by

$$\lambda_2 = g^{(1)}(\hat{N}_1)g^{(1)}(\hat{N}_2), \qquad (2.13)$$

and in general if there is p-point cycle, $\hat{N}_1, \hat{N}_2, \ldots, \hat{N}_p$, the stability-determining eigenvalue, λ_p, is given by

$$\lambda_p = g^{(1)}(\hat{N}_1)g^{(1)}(\hat{N}_2)\cdots g^{(1)}(\hat{N}_p). \qquad (2.14)$$

Applying equation 2.13 to the LH_1 population, we conclude that $\lambda_2 = 0.75$ for the linear logistic model and $\lambda_2 = 0.84$ for the exponential model. Thus, both models predict stable two-point cycles.

In contrast to the discrete-time models discussed here, the simple continuous-time models we have considered do not produce cycles or chaos. This is essentially because in these models it is assumed that the population can adjust instantly to the current density conditions (i.e., there is no time lag in the negative-feedback mechanism), an assumption that might not hold for many biological populations. Continuous-time models can, however, be modified such that the current rates of reproduction are determined by the density conditions that existed T time units in the past. With this type of time delay, we find more complex behavior-like cycles in continuous-time models (Cushing, 1977; Nisbet and Gurney, 1982; Renshaw, 1991, pp. 88–93; Hastings, 1997, p. 92). The discrete-time models, in fact, have these types of time delays built into them since reproduction is determined by the density of progeny produced one time unit ago.

Chaos

The next important question we tackle is how one characterizes chaotic population dynamics. The very word *chaos* suggests a lack of any structure, and this is true to some extent.

For instance, the variation in population size produced by a chaotic population superficially appears similar to random noise. Yet, there is often a precise set of deterministic equations that drive the dynamics of chaotic populations, and the behavior of these equations is not random. One of the hallmarks of chaotic dynamics is extreme sensitivity to initial conditions, implying that the trajectories of two populations that initially start very close to each other will, over time, diverge and become increasingly different. From this definition, it is clear that two trajectories that start from different points cannot intersect each other if their dynamics are chaotic; otherwise, the paths would be identical from the time of intersection. With discrete-time models, two different trajectories may cross but never intersect each other. With the simple continuous-time models considered here, it is impossible for two continuous trajectories to cross paths without intersecting at one time point. For this reason, it is more difficult for continuous differential equations to show chaotic behavior. In fact, continuous nonlinear equations do not exhibit chaos until there are at least three or more independent variables (species, genotypes, etc.). Discrete-time models are not so constrained and may exhibit chaos in one dimension (if the nonlinear function is not invertible) or two.

The study of nonlinear systems that give rise to chaos has also identified some unifying processes at work. Assume that there is a single variable that determines the stability of a nonlinear equation, such as the logistic. Let r_1 be the value of that parameter where the stable-point equilibrium bifurcates to a period-2 equilibrium. Likewise, r_2 is the parameter value at which the two-point cycle gives way to a four-point cycle and so on. We then define delta n as the ratio $\delta_n = (r_n - r_{n-1})/r_{n+1} - r_n$. The Feigenbaum delta is then defined as $n \xrightarrow{\lim \delta_n} \infty = 4.66920161\ldots$ This result is independent of the particular nonlinear function that gives rise to these cycles. In physical systems in which the δ_n can be reasonably estimated, there is general agreement with the Feigenbaum delta. This result suggests a unifying structure to period-doubling phenomena. Consequently, there is probably little practical application of the Feigenbaum delta to problems in ecology. For most biological populations, it is extremely difficult to determine the precise conditions under which a two-point cycle would give way to a four point cycle, for instance, and therefore to empirically estimate r_n.

The stability of nonlinear models can be determined by a combination of one or more parameters that we call the control parameters. As the value of the control parameter varies, the behavior of the model may change until it exhibits chaos. There are a variety of routes to chaos that different models may display (Hilborn, 1994). In ecological models, three routes have been seen: (1) period doubling, (2) quasi-periodicity, and (3) intermittency (Ruxton and Rohani, 1998). We have already discussed the period-doubling route, which is one of the most common routes to chaos in ecological models. Quasi-periodicity refers to periodic oscillations that are influenced by two or more periods for which the ratio of the two frequencies is not a rational number. This gives rise to population trajectories that look as though they repeat but in fact do not. As the control parameter is varied, the system moves from this quasi-periodic behavior to chaos. The presence of multiple-frequency oscillations can be detected by the use of time-series analysis (reviewed in chapter 3). Several host-parasite models exhibit this form of chaos (Rohani et al., 1994; Rohani and Miramontes, 1995). Finally, intermittency refers to trajectories that show irregularly occurring periods of chaos separated by durations of periodic behavior. As the control parameter is varied, the relative duration of the chaotic episodes becomes longer until the behavior is always chaotic, with no intervening durations of periodic dynamics. Such dynamic behavior has been observed in models of two genotypes with different population dynamic parameters (Doebeli, 1994), and in host-parasite models in which the host has three different phenotypic classes (Cavalieri and Koçak, 1995). We next review one of the characteristic indicators of chaos.

Suppose our population-dynamic equation predicts a series of population sizes (an orbit) that look like N_0, $N_1, \ldots, N_k$. If we started at a slightly different point, say $N_0 + y_0$, would the trajectories depart from the previous orbit or stay close to it? We can answer this question by looking at the product of the partial derivatives evaluated at the original orbit, in a manner similar to our previous stability analysis:

$$y_{k+1} = g^{(1)}(N_k)g^{(1)}(N_{k-1}) \cdots g^{(1)}(N_0)y_0 \qquad (2.15)$$
$$= \left[g^{(1)}(N_0)\right]_{k+1} y_0$$

Whether the initial perturbation, y_0, is growing or shrinking can be assessed by looking at $|y_{k+1}|/|y_0|$. Formally, the determination of chaotic dynamics rests on evaluating the Lyapunov exponents (Ott, 1993; Ellner and Turchin, 1995). For the first-order difference equations discussed so far, there is only one Lyapunov exponent, although for higher dimensional systems the number of Lyapunov exponents may be as high as the dimensionality of the system. Because the value of the initial perturbation is arbitrary, let us assume that $y_0 = 1$ for the following example. The Lyapunov exponent, $h(N_0)$, is then defined as follows:

$$\lim_{k \to \infty} \frac{1}{k} \ln(|y_k|) = \lim_{k \to \infty} \frac{1}{k} \ln \left(\left| \left[g^{(1)}(N_0)\right]_k \right| \right)$$

Lyapunov exponents that are positive characterize chaotic systems. This rule implies that the geometric mean of the first derivatives in equation 2.15 is greater than 1, and thus that the perturbation from N_0 is growing. Of several methods for numerically estimating Lyapunov exponents, we have used one described by Ott (1993) to estimate the Lyapunov exponents for the linear logistic (2.1) and exponential (2.3) models as a function of a_1 (fig. 2.4). The results emphasize

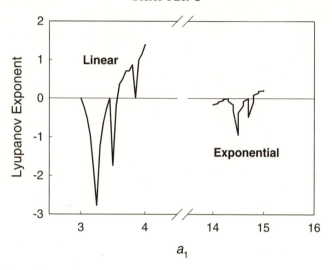

FIG. 2.4. Lyapunov exponent for the linear (2.1) and exponential (2.3) models of population growth as a function of the parameter a_1. Values for a_2 were taken from the LH_1 population shown in table 2.2. Each estimated Lyapunov exponent was based on the population sizes following from $N_0 = 650$ and $y_0 = 1$. A total of nine new generations of population sizes were generated and y_9 estimated from equation 2.15. The first estimated Lyapunov exponent was then taken as $h_1 = y_9/10$, y_9 was set back to 1, and the process was repeated starting at N_{10}. The final estimate of the Lyapunov exponent was based on the average of 1000 sequential values of h_1.

that even when the linear logistic and exponential models are producing cycles, the Lyapunov exponent will be negative if these cycles are stable.

SECOND- AND HIGHER-ORDER MODELS

This chapter started out with the simplest formulation of the discrete-time models, one that assumed that population size depends only on the size of the population in the very last generation or time interval. Models that depend on the most recent population size are called first order. If population size depends on two previous population sizes, it is second order and so on. There are several important biological phenomena that cause current population size to depend on population size in several previous generations. Here we consider two such phenomena: age structure and the dependence of adult fertility on pre-adult density. In theory, these phenomena may have an important impact on population stability. More important, if observations are made on populations in which these phenomena are important, failing to take this factor into account may give misleading inferences about the stability of populations. We develop these ideas by first discussing populations with age structure.

Age Structure

If adults survive from one time interval to the next and remain capable of reproduction, then we are dealing with an age-structured adult population. This factor increases the complexity of any model attempting to capture the dynamics of such a population. To illustrate some of the problems generated by age structure, we consider a simple model. We assume that two adult age classes with population sizes N_{1t}, N_{2t}. Individuals of both age classes are assumed to be capable of reproduction. If the total adult population size is $N_t(=N_{1t} + N_{2t})$, age-class transitions are given by the following equations:

$$N_{1,t+1} = N_t(a_1 + a_2 N_t) \qquad (2.16a)$$

$$N_{2,t+1} = b N_{1,t} \qquad (2.16b)$$

Thus, the number entering the first age class depends on the previous total adult population size in a density-dependent fashion. Survivorship from the first adult age class to the second is density independent and is determined by the survivorship probability, b. To emphasize the dependence on population sizes for more than one generation, the recursion for $N_{1,t+1}$ can be rewritten as follows:

$$N_{1,t+1} = \left(N_{1,t} + b N_{1,t-1}\right)\left(a_1 + a_2 N_{1,t} + a_2 b N_{1,t-1}\right) \quad (2.17)$$

When $b = 0$, equation 2.16 reduces to the linear logistic model (2.1). The equilibrium adult population sizes for equation 2.16 are as follows:

$$\hat{N}_1 = \frac{1 - a_1(1 + b)}{a_2(1 + b)^2} \qquad (2.18a)$$

$$\hat{N}_2 = b\hat{N}_1 \qquad (2.18b)$$

Box C shows the general method for the analysis of the stability of systems of difference equations. For model 2.16, we have ascertained the stability-determining eigenvalue for a range of b values, using a_1 and a_2 values from the LH$_1$ population in table 2.2 (fig. 2.5).

The example in figure 2.5 shows that age structure may, under some circumstances, stabilize a cycling population. Interestingly, in this example, the stability of a single-point equilibrium is consistent with intermediate values of survival from the first age class to the second. Clearly, age structure is an important detail that needs to be taken into account in empirically assessing the stability of a population's dynamics. Our qualitative view of population stability may be substantially altered when age structure is included in population-dynamics models. It has, however, been more difficult to draw any specific conclusion about the effects of age structure on population stability that would apply across a broad spectrum of biologically relevant situations. For instance, Guckenheimer et al. (1977) concluded that "general 'rule of thumb' appears to be that as the dimensionality of the system increases the amount of nonlinearity required to produce complex behavior decreases." In contrast, Charlesworth (1994) provided evidence supporting the notion that age-structured populations are more likely to exhibit stable behavior. However, Charlesworth also noted that as the prereproductive period lengthens, this stabilizing effect of age structure declines. Swick's (1981) position was intermediate, maintaining that age structure makes it less

---

BOX C

STABILITY ANALYSIS FOR SYSTEMS OF NONLINEAR EQUATIONS:

We consider vector-valued population size data (e.g., age classes) with $\mathbf{N}_t = (N_1, N_2, \ldots, N_d)^T$, where the superscript T denotes a matrix transpose and d is the total number of different age classes. The transition of this vector from one time interval to the next is governed by

$$N_{1, t+1} = g_1(\mathbf{N}_t),$$
$$N_{2, t+1} = g_2(\mathbf{N}_t),$$
$$\cdot$$
$$\quad\quad\quad\quad\quad\quad\quad\quad\quad\quad (2.19)$$
$$\cdot$$
$$N_{d, t+1} = g_d(\mathbf{N}_t)$$

---

likely to observe higher-order cycles or chaos, but perhaps more likely to induce simple cycles.

All said, it is clear that when one is assessing data from real populations, attention must be paid to the issue of age structure. For example, suppose a population has age structure of the sort described by equations 2.16, and that estimates of total population size at different times are collected from this population and are used to evaluate the dynamics of the population. If the population is treated as if it did not have age structure, will the correct conclusion regarding population stability still be reached?

To address this question we simulated a series of 20 generations of population growth for the model with two age classes (2.16), adding environmental noise to each adult age class in a manner following Dennis et al. (1995):

$$\ln[N_{1, t+1}] = \ln[N_t(a_1 + a_2 N_t)] + \varepsilon_{1t} \quad (2.20a)$$

$$\ln[N_{2, t+1}] = \ln[b N_{1, t}] + \varepsilon_{2t} \quad (2.20b)$$

BOX C cont.

We assume that there is an equilibrium for this system given by $\hat{\mathbf{N}} = (\hat{N}_1, \ldots, \hat{N}_d)^T$. The stability of this equilibrium is determined by the $d \times d$, Jacobian matrix ($\mathbf{J}$), which contains the first derivatives of the functions in equation 2.19 evaluated at the equilibrium $\hat{\mathbf{N}}$:

$$\mathbf{J} = \begin{bmatrix} \frac{dg_1(\hat{\mathbf{N}})}{dN_1} & & \frac{dg_1(\hat{\mathbf{N}})}{dN_d} \\ & \ddots & \\ \frac{dg_d(\hat{\mathbf{N}})}{dN_1} & & \frac{dg_d(\hat{\mathbf{N}})}{dN_d} \end{bmatrix}$$

The system's stability depends on the modulus of the largest eigenvalue, $|\lambda^*|$, of $\mathbf{J}$. For real eigenvalues, the modulus is just the absolute value. For complex eigenvalues $(a + bi)$, the modulus is equal to $\sqrt{a^2 + b^2}$. If $|\lambda^*| < 1$, then the equilibrium, $\hat{\mathbf{N}}$, is stable. The eigenvalues of $\mathbf{J}$ are found by solving the equation, $\det(\mathbf{J} - \lambda\mathbf{I}) = 0$, where *det* stands for the determinant of a matrix.◆

The ε_{xt} were uncorrelated, and each was normally distributed with mean zero and variance 0.0025. The resulting total population sizes are shown in figure 2.6 for three different values of a_1, with a_2 and b being held constant. Next we applied one of several methods for estimating population stability (reviewed in chapter 3) to this simulated data set, assuming that a particular model describes the population dynamics of this system. Estimates of the model parameters were then made from the observed population size variation. These estimates were in turn used to estimate the stability-determining eigenvalue. This eigenvalue may give rise to misleading conclusions about population stability if either the parameter estimates are poor, or if the original model does not adequately describe the dynamics of the population.

We applied maximum-likelihood techniques to the data shown in figure 2.6 to estimate the parameters of model

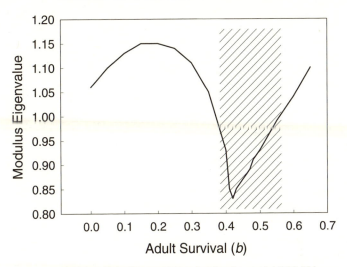

FIG. 2.5. Modulus of the largest eigenvalue for the model 2.16. Values for a_1 and a_2 were taken from fitting the linear model (2.1) to data from the LH_1 population (table 2.2). The value of the adult survival parameter b was allowed to vary. When $b = 0$, the eigenvalue is the same as in table 2.2 (-1.06). For $b > 0$, there is a shaded region where the equilibrium (equation 2.18) is stable. Thus, the addition of age structure can make a population that is in a two-point cycle settle down to a single equilibrium point if there is moderate survival from the first age class to the second.

2.17 by letting $N_{1,t} - N_t$. These estimates were made separately under two different assumptions. In the first case, we assumed that the dynamics were first order, so N_{t+1} depends only on N_t, and therefore $b = 0$. In the second case, we assumed that the dynamics were second order and estimated b directly from the observations. Note that this second-order model is still different from equation 2.16 since we are modeling changes only in total population size. We assumed that the different age classes are indistinguishable. Using the estimated parameters, we then obtained an estimated eigenvalue and compared it with the true eigenvalue of the deterministic system (equation 2.16). These results are shown in table 2.3.

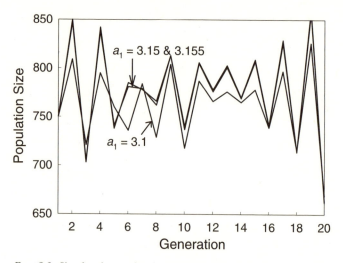

F<small>IG</small>. 2.6. Simulated growth of an age-structured population. Adult population $(N_{1t} + N_{2t})$ size from equations 2.20a and 2.20b are shown for three different values of a_1. The two curves for $a_1 = 3.15$ and 3.155 are close to each other over all 20 generations of data. The same set of random variables, ε_{xt}, was used in each of the three simulations. For each simulation, $a_2 = -0.00312$ and $b = 0.43$.

When a_1 is 3.1, the true stability-determining eigenvalue has modulus 0.85 (in fact, the eigenvalue is complex), and thus the system approaches a stable point. However, when the data are analyzed assuming a first-order difference equation (e.g., assuming $b = 0$) the estimated eigenvalue suggests that the equilibrium point will be unstable. The second-order model (table 2.3) does a much better job of estimating the sign and magnitude of the eigenvalue. Of course, we have considered only three examples in table 2.3. To reach a more general result would require a systematic and detailed examination of the two different estimation schemes used in table 2.3. However, these limited results clearly question the validity of ignoring the additional time dependence created by age structure.

Dennis and Taper (1994) hold out the hope that it may still be possible to model population dynamics or some

TABLE 2.3. Results of the analysis of population data in figure 2.6

		First Order		Second Order		
True	a_1	—	—	3.1	3.15	3.155
values	a_2	—	—	−0.00312	−0.00312	−0.00312
	b	—	—	0.43	0.43	0.43
	$|\lambda|$	—	—	**0.85**	**0.993**	**1.005**
	σ^2	—	—	0.0025	0.0025	0.0025
Estimated	$\hat{a}_1$	3.07	3.08	2.906	2.919	2.926
values	$\hat{a}_2$	−0.00271	−0.00266	−0.00225	−0.00222	−0.00224
	$\hat{b}$	0	0	0.207	0.206	0.189
	$\hat{\lambda}$	**1.07**	**1.08**	**0.987**	**1.0012**	**1.011**
	$\hat{\sigma}^2$	0.00055	0.00054	0.000455	0.000463	0.000472

*The three curves in figure 2.6 were analyzed assuming a first-order difference equation ($N_t = N_{t-1}(a_1 + a_2 N_{t-1})$) and a second-order difference equation ($N_t = (N_{t-1} + b N_{t-2})(a_1 + a_2 N_{t-1} + a_2 b N_{t-2})$) model. The first five rows show the parameter values used to generate the results in figure 2.6. The last five rows show estimates obtained for each of the two models. The actual eigenvalues (bold) for the deterministic model generating these data are shown along with the estimated eigenvalue (bold) derived from the maximum-likelihood estimates of the model parameters (indicated by the hats ˆ).

index of population size by simple first-order equations under appropriate conditions. Livdahl and Sugihara (1984), for example, described a method for estimating population growth rates from cohort data in age-structured populations. Their index is a function of female survivorship to reproductive age, the size of females, and the relationship between size and female fecundity. Their method typically applies only to growth rates for populations at a stable age distribution and in populations with high juvenile mortality, negligible adult mortality, and female fecundity that varies with adult size but not age. Similarly, Barlow's (1992) method for estimating population growth requires a stable size distribution. Barlow's index assumes fecundity to depend on adult size, and adult mortality to be age independent.

In variable environments or especially in populations that are cycling or chaotic, the assumptions of stable-age or size distributions required by Livdahl, Sugihara, and Barlow are dubious. Likewise, the assumptions about age-specific survival would limit the species and populations with which these techniques could be used. The point we wish to stress is that the effects of age structure on population dynamics must be considered carefully when trying to analyze data from populations in nature.

Pre-Adult Density Effects on Adult Reproduction

Even populations that have fully discrete generations and no age structure may not be properly modeled by first-order difference equations if there are particular kinds of interactions between different life stages. Prout and McChesney (1985) were the first to study this issue systematically. We briefly discuss some of the kinds of problems that such interactions between life stages may cause in analyzing population-dynamic data based on censusing only a single life stage. We return to this issue in detail in parts of the following chapters when we discuss specific model systems. For the present discussion, we focus on the kind of general life cycle considered by Prout and McChesney.

eggs   larval survival   adults $[n_t G(n_t)]$ eggs next generation

n_t $G(n_t)$ $n_t G(n_t)F(n_t)$

egg production

$F(n_t)$

FIG. 2.7. Discrete life cycle with two density-dependent life stages, larval survival, and female fecundity. Fecundity is assumed to be a function of the degree of larval crowding.

For this type of population, which we illustrate in figure 2.7, the following is a model of egg dynamics:

$$n_{t+1} = F(n_t)G(n_t)n_t. \tag{2.21}$$

However, if the census stage are adults, $N_t (= G(n_t)n_t = H(n_t))$, the recursion can only be reconstructed if the function $H(n_t)$ is invertible. For a variety of species, empirical observations suggest that this function is humped and, consequently, not invertible (Prout and McChesney, 1985). We have reproduced one set of empirical data collected by Rodriguez (1989) for *Drosophila melanogaster* (fig. 2.8). If a census of this *Drosophila* population shows 150 adults, there are two possible egg densities (N_L, N_U) that could each give rise to this number of adults. Adults raised at a larval density of N_L would be expected to be larger than adults raised at the larval density N_U. Because larger females lay more eggs, these differences in size have important consequences for rates of population growth:

$$N_{t+1} = G(n_{t+1})n_{t+1} = G\left[F(n_t)N_t\right]F(n_t)N_t \tag{2.22}$$

However, n_t depends on n_{t-1} and N_{t-1} and so on. For many organisms, the adults are the most conspicuous and easily sampled. Consequently, when adult numbers are counted but their fertility is a function of their pre-adult densities, then the recursion in adult numbers depends on many previous adult densities in a rather complicated fashion. If the underlying stability of a population cannot be determined

61

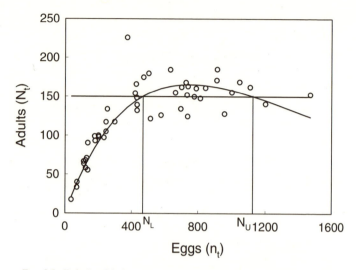

Fɪɢ. 2.8. Relationship between egg numbers and adult numbers within one generation of *Drosophila melanogaster* (from Rodriguez, 1989). The solid curved line is an exponential model that has been fit to these data. An adult population of 150 may have arisen from either an initial batch of N_L eggs or N_U eggs. The relationship between egg number and adult numbers is not one to one.

from the numbers of adults alone, then population studies will be complicated. Although, in principle, the adult population size may depend on many previous population sizes, practically it may be possible to get reliable estimates of population stability by examining only a few previous population sizes.

We examine this problem by investigating one model considered by Prout and McChesney (1985). The combination of a linear survival function and exponential fertility yields

$$n_{t+1} = \frac{1}{2} F \exp(-f n_t)(S - s n_t) n_t, \qquad (2.23)$$

where S is the maximum larval survival rate at low density, s reflects sensitivity of survival to larval crowding, F is the maximum fecundity at low density, and f measures the sensitivity of female fecundity to crowding. We have used the parameter estimates for these functions obtained by Prout and McChesney for *D. melanogaster*, and have estimated an equilibrium egg number (1758) and stability-determining eigenvalue (−1.25). We have used equation 2.23 to simulate 100 generations of adult population sizes with random noise (fig. 2.9).

The adult data in figure 2.9 have been used to estimate the parameters of the linear logistic model (2.1) and the second-order linear model:

$$N_t = N_{t-1}(a_1 + a_2 N_{t-1}) + N_{t-2}(b_1 + b_2 N_{t-2}) \qquad (2.24)$$

The equilibrium of equation 2.24 is $\hat{N} = (1 - a_1 - b_1)(a_2 + b_2)^{-1}$. The stability of this equilibrium can be assessed by the method described in box D.

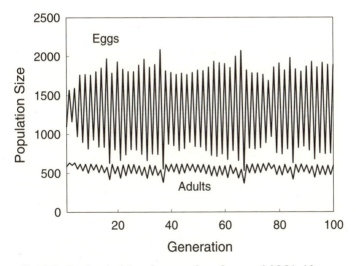

FIG. 2.9. Simulated adult and egg numbers, from model 2.24 with random environmental noise. The parameter estimates come from Prout and McChesney (1985) and were 0.845 (S), 0.00028 (s), 16.429 (F), and 0.001 (f).

63

BOX D
Stability Analysis for a kth-Order Nonlinear Difference Equation:

We consider a recursion for population size in which the present population size depends on the previous k-values of the population size:

$$N_t = H(N_{t-1}, N_{t-2}\ldots, N_{t-k}). \qquad (2.25)$$

Equation 2.25 can be solved for a point equilibrium by setting $N_t = N_{t-1} = \cdots = N_{t-k} = \hat{N}$. The stability of this equilibrium point can be evaluated using Taylor's series to approximate $\hat{N} + \varepsilon_t$:

$$\hat{N} + \varepsilon_t \cong \hat{N} + \varepsilon_{t-1} \frac{dH(\cdot)}{dN_{t-1}}\bigg|_{N_{t-1}=\hat{N}} + \cdots + \varepsilon_{t-k} \frac{dH(\cdot)}{dN_{t-k}}\bigg|_{N_{t-k}=\hat{N}}$$

If we let $\varphi_i = (dH(\cdot))/dN_{t-i}|_{N_{t-i}=\hat{N}}$, the perturbations to the equilibrium grow according to the linear-homogeneous difference equation:

$$\varepsilon_t = \varphi_1 \varepsilon_{t-1} + \cdots + \varphi_k \varepsilon_{t-k}$$

The maximum-likelihood estimates of the adult data in figure 2.9 were obtained for models 2.1 and 2.24. The eigenvalue for the first-order model is -0.73. Thus, with this model the analysis of the adult data suggests that the population should be stable. Results from the second-order model are quite different. The largest eigenvalue for model 2.24 is 1.29. In this case, the inclusion of an additional generation of adult numbers permits the correct evaluation of this population's stability. This simple example suggests that when analyzing the stability of populations, even those with fully discrete generations, the details of the effects of density dependence on various life stages can be critical.

BOX D cont.

The stability-determining eigenvalue is the largest among the possibly k-distinct solutions $(\lambda_1, \lambda_2, \ldots, \lambda_k)$ of the polynomial (Goldberg, 1958, pp. 169–171), $\varepsilon^k - \varphi_1 \varepsilon^{k-1} - \cdots - \varphi_{k-1} \varepsilon - \varphi_k = 0$. For the second-order linear model (2.24), the coefficients of the stability-determining quadratic equation are as follows:

$$\varphi_1 = a_1 + \frac{2a_2(1 - a_1 - b_1)}{a_2 + b_2}$$

$$\varphi_2 = b_1 + \frac{2b_2(1 - a_1 - b_1)}{a_2 + b_2} \quad \blacklozenge$$

EVOLUTION OF POPULATION STABILITY

The analysis of simple population dynamic models has revealed that population stability depends on various parameters that affect density-dependent rates of population growth. MacArthur and Wilson (1967) made the first serious attack on the problem of the evolution of population growth characteristics with the articulation of the theory of r- and K-selection. There has since been substantial progress in theoretical and experimental research in density-dependent natural selection (see Mueller, 1997, for a recent review). These ideas have suggested that evolution may mold the rates of population growth. Biological populations harbor genetic variation for traits affecting the value of population growth parameters that, in turn, can affect population stability (Mueller and Ayala, 1981c). Clearly, it is plausible to investigate the possibility that density-dependent natural selection may mold population stability just as it may mold population growth rates (Mueller and Ayala, 1981a; Mueller et al., 1991b).

Doebeli and de Jong (1999) point out that population stability is enhanced when genetic polymorphisms exist for certain population-dynamic parameters. Under this theory, stability is a by-product of genetic variability rather than a result of directional increases or decreases in life-history parameters that accompany natural selection.

In our view, population stability is more likely to be a by-product of individual life-history traits that are directly connected to genotypic fitness. Thus, natural selection may affect the evolution of fecundity, and this evolution may reflect the genetic correlations between fecundity and other life-history traits. A direct consequence of the evolution of female fecundity may be changes in population stability. This view contrasts with the view of the dynamical properties of a population as a trait that evolution may mold directly. Ferrière and Fox (1995), for example, speak about adaptive chaos and suggest that "chaos may be an easy way to generate variability and uncertainty." This theory suggests that a by-product of population dynamics is what drives evolution, whereas we feel that it is the fitness-related traits of individuals that are the focus of evolution, and that stability characteristics may be molded indirectly by such evolution. We now review the theory that has been developed in this general area.

Theories of the evolution of population stability have involved explanations based on mechanisms of both individual and group selection. The argument based on group selection is that unstable populations will more often have their population size reduced to small numbers (Thomas et al., 1980; Berryman and Millstein, 1989). During such a valley in population size, extinction may occur, perhaps partly due to enhanced susceptibility to environmental variation. To the extent that a species consists of many such populations that are essentially genetically isolated (otherwise there would be no genetic variation among populations), the environments that remain after population extinction may be recolonized by some neighboring population that is presumably more persistent. In the absence of empirical data supporting the special population structure needed to make this process work, it is difficult to take the group selection arguments very seriously. It is also reasonable to assume that environmental rather than genetic differences are often responsible for the relatively unstable dynamics of particular populations. In such cases, recolonization of a habitat patch following extinction may not represent any evolutionary change, being no more than an expression of migration of individuals from a habitat patch with an environment supporting relatively stable dynamics.

Allen et al. (1993) have stood this argument on its head, instead considering the extinction or persistence of sets of populations. They argue that if we consider a species consisting of many populations linked by low levels of migration, the chance of the species becoming extinct is reduced by chaos. This conclusion hinges on the notion that chaos produces uncorrelated variation in neighboring populations. Global noise, like weather, produces correlated variation among the subpopulations, with local environmental noise uncorrelated among subpopulations. Allen et al. suggest that chaos amplifies the heterogeneity of the local populations and thus reduces the likelihood of species-wide extinction. However, in the absence of this local environmental variation, chaos does result in increased local and species-wide extinction rates.

On the other hand, there are related, and perhaps more plausible, explanations for the evolution of population stability. For instance, populations that do undergo repeated bottlenecks due to population size fluctuation may experience increased levels of inbreeding. In outbred, highly fecund species, inbreeding may substantially reduce fecundity. Because population stability is often affected by maximum rates of population growth, which in turn depend on fecundity, inbreeding may indeed lead to enhanced stability for certain species. However, we expect this type of stability enhancement to be short lived if there is immigration from neighboring populations with high-fitness, outbred individuals.

67

The basic theory of density-dependent natural selection (Roughgarden, 1971) used the standard form of model 2.1 to describe population growth as follows:

$$N_{t+1} = N_t \left(1 + r - \frac{rN_t}{K} \right) \tag{2.26}$$

If genetic variation affects genotypic specific values of r and K, then in constant environments the outcome of selection depends on the relative population density. When population size is high, selection favors those genotypes with the highest K. If the population is kept at very low densities, then selection favors the genotypes with the highest values of r. It is a small step to move from the evolution of population growth rates to the evolution of population stability. Heckel and Roughgarden (1980) made this step by first suggesting that selection would favor reduced values of r in environments where K varied. This conclusion follows from the idea developed by Gillespie (1974) that natural selection favors a reduction in the variance in fitness. By decreasing r in variable environments, populations near their carrying capacity may achieve a reduction in the variance in fitness. Thus, for model 2.26, the results of Heckel and Roughgarden suggest that natural selection in a variable environment tends to increase the deterministic stability of the equilibrium.

Turelli and Petry (1980) considered a class of models that had the general form $N_{t+1} = N_t G[(N_t/K)^\theta]$, where the function $G(\cdot)$ assumed either a linear, exponential, or hyperbolic form. The parameter θ has provided a better description of population-dynamic observations for some organisms, and there appears to be genetic variation that affects its value (Mueller and Ayala, 1981c). Their models permitted environmental variation to affect the carrying capacity or density-independent growth rates (by multiplying $G(\cdot)$ by $1+z$ where z has mean zero and variance σ^2). They found that when the parameter r is allowed to evolve in these equations, stability may increase, decrease, or be unaffected. However, when θ

was allowed to evolve, more consistent results were observed and selection often resulted in population stability.

Turelli and Petry (1980) dealt with populations that initially had parameter values that produced stable dynamics. Mueller and Ayala (1981b), Stokes et al. (1988), and Gatto (1993) examined the evolution of stability in populations initially at a stable cycle or chaos. Typically for these models to cause populations to evolve stable dynamics, some type of trade-off is required in parameters of the population-dynamic models. For instance, Mueller and Ayala (1981b) showed that populations may evolve from a two-point cycle to a stable point if density-dependent viability trades off with fecundity. Thus, under these models there exist genotypes with increased viability but decreased fecundity. Nevertheless, these genotypes have sufficiently high fitness that they can replace the resident genotype responsible for the two-point cycle. Of course, this sort of genotype might be favored even if the population was not cycling. However, in most simple ecological models, the highest density in a two-point cycle exceeds the carrying capacity, and thus the strength of density-dependent selection changes. The decrease in fecundity that ensues ultimately stabilizes the population. Gatto (1993) also described conditions under which populations may evolve into chaos. However, fairly special combinations of life-history parameters or special population structure (Gomulkiewicz et al., 1999) are required for the evolution of chaos.

Hansen (1992) suggested that the model dependence of these results may be due to the manner in which the models are constructed. In some cases, such as model 2.27, the parameter that controls stability (r) is different from the parameter under direct selection at high density (K). In other models, this separation is not present. Hansen suggested that selection at low densities typically favors instability while the opposite occurs at high density.

The theoretical debate over the evolution of population stability has been aired recently (Ferrière and Fox, 1995;

Doebeli and Koella, 1996; Fox, 1996). Ferrière and Fox argued that, in principle, natural selection can favor the evolution of chaotic dynamics in populations, and that this possibility needs to be considered seriously. Doebeli and Koella (1995) suggest that their own modeling efforts support the notion that selection is more likely to favor the evolution of stable rather than chaotic dynamics. Both these theories have not really clarified any of the issues raised by the previous theory. For instance, neither theory considers more than one functional form of population dynamics, despite Turelli and Petry's (1980) demonstration of model sensitivity. Special assumptions about the relationship of population parameters are ultimately critical to the evolution of stability in both models (Ferrière and Fox, 1995; Doebeli and Koella, 1995).

A major difficulty with assessing the various predictions of these models is that they depend on assumptions that can be evaluated only from empirical studies. For example, to what extent are pre-adult survival and fecundity correlated, and can one alter θ without changing r or K? Some of these issues could be more reasonably assessed if the population-dynamic models incorporated specific details of important life-history events of organisms (Christiansen, 1984). An important theme in our chapters on model systems is the use of models that specifically incorporate important life-history phenomena such as cannibalism in *Tribolium* or scramble competition for food in *Drosophila*. These models are to some extent less general than the simple models just discussed, but they are far more useful for the evaluation and design of experiments to critically assess predictions from theory.

Techniques for Assessing Population Stability

In this chapter, we explore how the techniques for determining the stability of models, reviewed in chapter 2, can be applied to data from real populations. There are several approaches to this problem, each with different strengths and weaknesses. One technique uses direct observations of population growth rates to estimate linear population dynamics in the vicinity of an equilibrium. This approach is obviously motivated from the mathematical definitions of stability reviewed in chapter 2. Another technique for assessing stability is based on evaluating the time-dependent behavior of population growth and using these results to infer the deterministic behavior of the population. This technique uses the tools of time-series analysis.

There are no formal distinctions between techniques that can be used with laboratory populations and those that can be used with natural populations. However, we typically have much more information about the factors controlling population growth in laboratory populations. Consequently, techniques based on specific models are more often applied to laboratory populations. Nevertheless, we find that some techniques, such as time-series analysis, are used with both laboratory and natural populations. Certain techniques are useful for distinguishing chaos from other dynamics but do not permit us to dissect stable points from stable cycles, whereas other techniques do not specifically identify chaotic dynamics but do differentiate a single stable point from other types of dynamic behavior. Although many techniques focus on the stability of the deterministic growth process,

71

others yield stability estimates for the deterministic and stochastic components of population growth.

LINEARIZED POPULATION DYNAMICS
IN THE VICINITY OF AN EQUILIBRIUM

We saw in chapter 2 that a Taylor series could be used to provide a linear approximation of the dynamics of a population in the vicinity of an equilibrium point. In principle, if one could collect empirical estimates of rates of population growth in the vicinity of the carrying capacity, these could be used to estimate the linear dynamics directly. The potential advantage of this technique is that one need not assume that any particular nonlinear model appropriately describes the processes underlying the growth of the population. Growth rates of laboratory populations may be collected by properly designed experiments over a single generation. Observations collected in this fashion and their stability estimates can then be compared to the time-dependent behavior of independent populations maintained over many generations. There are several drawbacks to this technique, however. For non-laboratory populations, it is rarely possible to collect observations of density-dependent rates of population growth around the carrying capacity. There is also the difficult question of practically defining the region about the carrying capacity in which dynamics are expected to be approximately linear. If one chooses a range of densities that are too close to each other, then experimental error may preclude any accurate estimation of the slope of the linear dynamics. If, however, the range of densities chosen is too large, then the dynamics are unlikely to be linear.

To our knowledge, this technique has been used only once (Mueller and Ayala, 1981b), on laboratory populations maintained by a technique called the serial transfer system (Ayala, 1965a) in which an adult breeding population is maintained with overlapping generations. Because it is a fairly complicated technique and has been used in several other studies that are discussed later, we have outlined the basic steps of the serial transfer system in figure 3.1. An adult census is made at regular (usually one-week) intervals, and age-class numbers are unknown. The total number of adults at the census, N_t, is composed of surviving adults from the previous week, $g_1(N_{t-1})$, and adults who have emerged over the last week from bottles that are two $(g_2(N_{t-2}))$, three $(g_3(N_{t-3}))$, and four weeks old $(g_4(N_{t-4}))$, respectively (fig. 3.1). The number of cultures maintained may be different than four, depending on the species of *Drosophila* used.

The model shown in figure 3.1 presumes that recruitment from old cultures is dominated by the density of adults that originally laid eggs in that culture, and is also essentially independent of the age structure of the population. Justification for this untested assumption is that in standard laboratory populations, the average life span is short (probably two weeks or less), and larval mortality is high due to severe crowding and overproduction of eggs. Thus, even though newly emerged adults lay eggs in their larval habitat (e.g., cultures $t - 2$, $t - 3$, and $t - 4$ in figure 3.1), these eggs almost never successfully develop and emerge before the culture is discarded. Figure 3.2 shows the adult population size variation for two populations maintained by the serial transfer system for 38 weeks.

Rates of population growth at any desired density can be estimated for the serial transfer system using single-generation experiments, in which adults at a specified density, N^*, are placed in a single bottle for one week. At the end of the week, the number of survivors provides an estimate of $g_1(N^*)$. The number of emerging adults from this same bottle is counted one week later and provides an estimate of $g_2(N^*)$. Emerging adults are collected and counted at weekly intervals for the next two weeks, providing estimates of $g_3(N^*)$ and $g_4(N^*)$. This type of experiment can be repeated at the same density multiple times and over a large

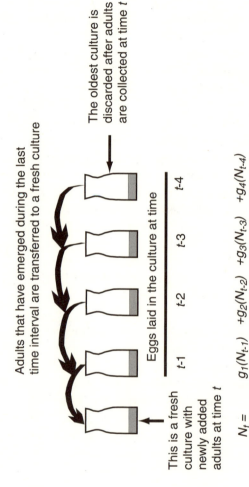

This is a fresh
culture with
newly added
adults at time t

Adults that have emerged during the last
time interval are transferred to a fresh culture

The oldest culture is
discarded after adults
are collected at time t

Eggs laid in the culture at time

| $t-1$ | $t-2$ | $t-3$ | $t-4$ |

$$N_t = \quad g_1(N_{t-1}) \quad +g_2(N_{t-2}) \quad +g_3(N_{t-3}) \quad +g_4(N_{t-4})$$

Fig. 3.1. Serial transfer system used to maintain populations of *Drosophila* with overlapping generations (after Mueller and Ayala, 1981c). The entire population consists of four cultures that have had eggs laid at different times. At regular (usually one-week) intervals, adults are collected from all four cultures making up the population (cultures with arrows above show the movement of adults). These adults are added to a fresh culture, where they lay eggs for the next week, while the oldest culture is discarded. In principle, there is a different nonlinear function for each culture, $g_i(N_{t-i})$, describing the number of adults that emerge as a function of the number of adults that laid eggs in that culture i weeks ago.

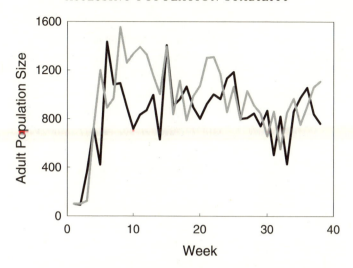

FIG. 3.2. Two populations of *Drosophila melanogaster* maintained in continuous culture by the serial transfer system. Each population was started with 100 adults. The black line is population 8, and the gray line is population 14 studied previously by Mueller and Ayala (1981b). Each population had been made homozygous for a whole second chromosome sampled from nature. Using data after the fifth week, the average size of population 8 is 912 ± 72 (95% confidence interval), and the average size of population 14 is 1032±80. These data are also given in table 3.1A of the appendix.

range of densities. For the two populations in figure 3.2, stability estimates were made from observations of survival and progeny production with N^* at 750 and 1000 adults for populations 8 and 14, respectively. These two densities were chosen because they were thought to bracket the carrying capacities of the populations. This assumption appears to have been accurate for population 8 but slightly off for population 14. Nevertheless, from these single-generation experiments, we can derive estimates of the carrying capacity for populations 8 and 14 that are independent of the observations in figure 3.2. (Raw data from the single-generation experiments for populations 8 and 14 are in table 3.2A.) These estimates predicted that the carrying capacity for population 8 was 880 and for population 14 was 990 (Mueller and Ayala, 1981c). Both of these estimates are well within the confidence intervals for the continuously cultured populations in figure 3.2. It is heartening to find that the single-generation experiments are capable of reasonably predicting the equilibrium population size of the continuously cultured population. Additional details of these experimental protocols are discussed in Mueller and Ayala (1981c) and Prout and McChesney (1985).

From the single-generation experiments, a linear model approximates the nonlinear functions in figure 3.1, $g_i(N_t) = a_{0i} + a_{1i}N_t$. The resulting model for N_t is a fourth-order, linear nonhomogeneous equation. Its eigenvalues are determined by methods described in box A.

Although Mueller and Ayala (1981b) appear to be the only ones to have used this linearization procedure to estimate population stability, several other studies have used rates of population growth from the serial transfer system to study population stability and other problems (Thomas et al., 1980; Hastings et al., 1981; Phillippi et al., 1987). In these studies, population growth rates were empirically determined using the single-generation experiments previously described. However, these studies assumed that population dynamics in the serial transfer system could be described by a first-order difference equation, $N_t = h(N_{t-1})$. The function $h(N_t)$ was then estimated from the single-generation experiments described earlier by letting $h(N^*) = \sum_i g_i(N^*)$. This summation is sometimes called the total productivity, since it represents the sum of all survivors and progeny. This model ignores the complicated time dependence of egg laying in the serial transfer system and would not be expected to yield estimates of growth rates that are relevant to populations maintained by the serial transfer technique. In fact, it is not clear if there is any population whose growth rates are estimated by $h(N^*)$.

BOX A
Stability of the Serial Transfer System:

When the functions in figure 3.1 are linear, the population growth model is $N_t = A + a_{11}N_{t-1} + a_{12}N_{t-2} + a_{13}N_{t-3} + a_{14}N_{t-4}$, where $A = \sum_i a_{0i}$. The stability-determining eigenvalues for this fourth-order, nonhomogeneous difference equation are the roots of the polynomial $\varepsilon^4 - a_{11}\varepsilon^3 - a_{12}\varepsilon^2 - a_{13}\varepsilon - a_{14} = 0$. Using the six independent observations of survivorship and progeny production at two densities for population 8 (see table 3.2A in the appendix), estimates of the four coefficients, a_{11}, a_{12}, a_{13}, and a_{14}, are $-0.476, 0.265, -0.122$, and -0.0827, respectively. The roots of the polynomial, in order of magnitude, are $-0.83, 0.36 + 0.39i, 0.36 - 0.39i$, and -0.36. These results may be obtained numerically from commercially available software such as Mathcad or Mathematica, or from numerical routines such as Laguerre's method (Press et al., 1986). Thus, population 8 should have a stable equilibrium, although the approach to equilibrium will be oscillatory. Depending on the initial conditions, the linear dynamics will be affected by all four eigenvalues, with the largest dominating asymptotically. The negative and complex eigenvalues will contribute to the oscillatory approach to equilibrium. If we let λ_3 and λ_4 be the real eigenvalues and $a \pm bi$ be the eigenvalues that form a complex conjugate, the general solution to the nonhomogenous linear equation just described is

$$N_t = C_1 r^t \cos(t\theta + C_2) + C_3 \lambda_3^t + C_4 \lambda_4^t + \tilde{N},$$

MODEL-BASED ESTIMATES OF STABILITY

The technique that we now consider is based on the use of specific nonlinear models to infer stability. We consider two major variations of this technique. The first variation assumes that a particular model provides a proper description of population dynamics for a given system and then uses observations from populations to estimate the model

Box A cont.

where the C_i's are constants determined from initial conditions, $\tilde{N}$ is a particular solution to the nonhomogeneous equation, r is the modulus of the complex eigenvalues, and θ is the solution of $\sin(\theta) = b/r$ and $\cos(\theta) = a/r$ (Goldberg, 1958, pp. 163–164). Thus, the cosine function contributes to the oscillatory approach to the equilibrium with a frequency θ. For population 8, $\theta = 0.825$ radians (0.13 cycles). For population 14, the polynomial coefficients a_{11}, a_{12}, a_{13}, and a_{14} are 0.23, -0.41, 0.0027, and 0.33, respectively. The roots, in order of magnitude, are $0.075 + 0.89i, 0.075 - 0.89i, 0.68$, and -0.60. The largest eigenvalue is complex with a modulus of 0.89. Thus, the equilibrium of population 14 should also be stable, although the approach to equilibrium will be oscillatory with a frequency due to the complex roots of 1.49 radians (0.236 cycles).

The largest eigenvalues derived here are slightly smaller than the values reported in Mueller and Ayala (1981b) as a result of a different estimation procedure. Mueller and Ayala used the jackknife technique and obtained values of 0.96 and 0.93 for the modulus of population 8 and the largest eigenvalue of population 14, respectively. The jackknife technique, which is related to the bootstrap, is a numerically intensive technique that can potentially reduce bias (Miller, 1974; Efron and Tibshirani, 1993). The difference in estimates observed here is a reflection of this bias. In practice, we recommend using either the jackknife or the bootstrap to estimate the largest eigenvalue. These procedures can be used to reduce bias and construct confidence intervals around the final estimates.◆

parameters. The second variation presumes ignorance of the appropriate model and uses observations from biological populations to determine the best model and estimate its parameters.

In both cases, once a population model is chosen and its parameters estimated, then the stability of the resulting equilibria can be determined by the techniques outlined in chapter 2 or by numerical techniques. Usually, numerical techniques are used when the model is sufficiently complicated to defy simple analysis of the equilibria. The advantage of the first approach is that the observations from the populations of interest are used only to provide estimates of model parameters. The uncertainty in these estimates can be estimated readily, and thus the uncertainty in the final conclusions is readily quantified. Of course the reliability of this method depends on how well the original model describes the dynamics of the system under study. Except for laboratory populations, there are few populations for which there is great certainty about the appropriate growth model. For most populations, the observations are used to determine the best model and estimate stability.

Models Chosen a Priori

The first serious attempt to assess the stability characteristics of natural populations was a survey of published data by Hassell et al. (1976). In this study, stability was assessed through the magnitude of the parameters of a single population growth model:

$$N_{t+1} = \lambda N_t (1 + aN_t)^{-\beta} \qquad (3.1)$$

For this model, the equilibrium population size is given by

$$\hat{N} = \frac{\exp[\ln(\lambda)\beta^{-1}] - 1}{a},$$

and the stability-determining eigenvalue is

$$1 - \beta \exp[\ln(\lambda)\beta^{-1}]\lambda^{-1/\beta}.$$

Hassell et al. use a variety of ad hoc techniques to estimate the parameters of equation 3.1. For instance, observations of maximum female fecundity were used to determine λ. The rate of population growth, according to equation 3.1, should be equal to λ at low density. However, weak motivation for the use of equation 3.1 with these data make it unlikely that fecundity data will provide accurate estimates of growth rates (λ) at low density. Hassell et al. (1976) also use a log transformation of equation 3.1 so that linear regression techniques may be used to estimate β. This procedure does not yield the same estimate of β as nonlinear regression on the untransformed data. Finally, the qualitative assessment of stability may depend critically on the precise model used. Morris (1990), who also reanalyzed some of the data in Hassell et al. (1976), showed that the use of standard nonlinear regression techniques and different growth rate models significantly affects the results of this type of analysis.

Rodriguez (1989) took a substantially different approach to the analysis of population stability. Rodriguez studied laboratory populations of *Drosophila melanogaster,* kept on a fully discrete regime of reproduction without age structure. The life cycle was separated into pre-adult survival and female fecundity. Survival ($V(n_t)$) was assumed to be a function of egg density (n_t), while female fecundity was a function of both egg density and adult density ($R(N_t, n_t)$) (see chapter 6 for more details). Rodriguez then estimated the parameters of the survival portions of this model from direct experiments in which eggs were crowded at different densities and the number of survivors counted. In a similar fashion, separate experiments were used to estimate the fecundity of females raised at different egg densities and cultured at different adult densities. The resulting model of egg density (n_t,) dynamics can be written as

$$n_t = n_{t-1} V(n_{t-1}) \tfrac{1}{2} R(N_{t-1}, n_{t-1}), \qquad (3.2)$$

where egg-to-adult survival is modeled by an exponential function,

$$V(n_t) = \exp(S - sn_t),$$

female fecundity, which is a function of egg and adult density (N_t), is given by,

$$R(N_t, n_t) = \exp(F - f_1 N_t - f_2 n_t),$$

and adult numbers are $N_t = n_t V(n_t)$. The stability-determining eigenvalue of equation 3.2 is given by

$$\lambda = 1 - s\hat{n} - \hat{n}[f_1 \exp(S - s\hat{n})(1 - s\hat{n}) + f_2], \qquad (3.3)$$

where $\hat{n}$ is the equilibrium egg number obtained from equation 3.2. The parameter estimates were substituted into equation 3.3 and yielded 0.064 as an estimate of the stability-determining eigenvalue, suggesting a stable equilibrium with a smooth approach to equilibrium. In arriving at this result, Rodriguez used only observations from parts of the life cycle. These parts were then reconstituted through equation 3.2 to complete an entire generation. This approach to estimating population dynamics is similar to the analysis of fitness components in population genetics (Prout 1965, 1971a, 1971b).

Rodriguez also collected nine generations of total egg numbers from 25 replicate populations but did not use them for parameter estimation. However, Turchin (1991) used these data to estimate population dynamic parameters, using a general first-order difference equation that is described in more detail in the next section (*Models Estimated from Data*). The stability-determining eigenvalue was −0.59 for this model. Time-series analysis of Rodriguez's data suggested a possible damped oscillation toward an equilibrium point. The eigenvalue obtained by Turchin is consistent with this result. However, is Turchins result at odds with the eigenvalue estimated by Rodriguez?

The difference between the results of Turchin and Rodriguez could be due to (1) the different models used in the analysis, (2) the different data used to estimate parameters, or (3) both of these factors. The second choice is probably important because the technique employed by Rodriguez makes some assumptions that may be violated in the running populations analyzed by Turchin. For instance, Rodriguez assumed that the larval and adult density act independently on fecundity. It may not be the case that the effects of 500 small adults on female fecundity are the same as those of 500 large adults, in which case this independence assumption is violated. Taken separately, these factors may have been small but together could be enough to account for the quantitative differences observed in the estimated eigenvalue.

It is also plausible that a confidence interval on the eigenvalue estimated by Rodriguez would include negative values. There are several possible ways to place a confidence interval on the eigenvalue given by equation 3.3. One simple method that uses a Taylor series approximation is called the delta method (Kendall and Stuart, 1969). Applying the delta method to the estimated value of equation 3.3 suggests that the 95% confidence interval is 0.064 ± 0.18 (see box B). Consequently, Rodriguez's results are not inconsistent with a negative eigenvalue.

As a last example, we consider several papers that have used different models to estimate the stability of populations of *D. melanogaster* maintained by the serial transfer system (Thomas et al., 1980; Hastings et al., 1981; Mueller and Ayala, 1981b; Philippi et al., 1987). As discussed earlier, several studies have assumed that the serial transfer system can be modeled by a first-order difference equation. When this is done, the net productivity statistic (= productivity − starting density) dramatically overestimates growth rates at low density. For instance, when productivity is used to estimate growth rates for populations 8 and 14 in figure 3.2 and the results are fit to the logistic equation, the estimates of r are 14.0 and 15.0, respectively. Thus, the stability-determining eigenvalues are 13 and 14 (compare to the direct estimates of the eigenvalues in box A). This would suggest chaotic population dynamics, a conclusion that is not supported or remotely suggested by any other analysis

---

BOX B

Delta Method:

The basic problem is to estimate the variance of a complicated function of a random variable or vector. Suppose we need to estimate the quantity M, which is a complicated function of k parameters, $c_1, c_2, \cdots, c_k$, as in $M = F(c_1, c_2, \ldots, c_k)$. We assume that we can obtain estimates of the parameters $\hat{c}_1, \hat{c}_2$, etc. These in turn are used to estimate M, so $\hat{M} = F(\hat{c}_1, \hat{c}_2, \ldots, \hat{c}_k)$. The function $F(\cdot)$ can be approximated by a Taylor series by expanding the function around the expected value of the parameters:

$$\hat{M} \cong F[E(c_1), E(c_2), \ldots, E(c_k)] + (\hat{c}_1 - E(c_1)) \quad (3.4)$$

$$\times \frac{dF}{dc_1}\Big|_{c_1 = E(c_1)} + \cdots + (\hat{c}_k - E(c_k))\frac{dF}{dc_k}\Big|_{c_k = E(c_k)}$$

Noting that $Var(\hat{M}) = E\big[\big(\hat{M} - F[E(c_1), \ldots, E(c_k)]\big)^2\big]$, we get the following estimate of variance:

$$\sum_i Var(\hat{c}_i)\left(\frac{dF}{dc_i}\right)^2 + \sum_i \sum_{j \neq 1} Cov(\hat{c}_i, \hat{c}_j)\frac{dF}{dc_i}\frac{dF}{dc_j} \quad (3.5)$$

In practice, we replace $E(c_i)$ with $\hat{c}_i$ when we finally estimate $Var(\hat{M})$. We now apply this technique to the eigenvalue in equation 3.3. The two experiments used to estimate S and s were different from those used to estimate f_1 and f_9. As a result, estimates of S and s were independent of f_1 and f_2. Applying equation 3.5 to equation 3.3 results in

---

of these data. Nevertheless, Hastings et al. (1981) developed a boundary layer model of population dynamics that would produce very high growth rates at low density but still exhibit stable dynamics about the carrying capacity. In effect, the model of Hastings et al. was motivated by an incorrect analysis of the experimental data rather than a novel biological phenomenon. In these examples, problems with the analysis lie not in the specific models chosen but in the techniques

BOX B cont.

$$Var(\hat{\lambda}) = Var(\hat{S})\phi_1^2 + Var(\hat{s})\phi_2^2 + Var(\hat{f}_1)\phi_3^2 \quad (3.6)$$
$$+ Var(\hat{f}_2)\phi_4^2 + 2Cov(\hat{S}\hat{s})\phi_1\phi_2$$
$$+ 2Cov(\hat{f}_1\hat{f}_2)\phi_3\phi_4,$$

where

$$\phi_1 = \frac{d\lambda}{dS} = -\hat{n}\hat{f}_1(1 - \hat{s}\hat{n})\exp(\hat{S} - \hat{s}\hat{n}),$$

$$\phi_2 = \frac{d\lambda}{ds} = -\hat{n} + \hat{n}^2 f_1 \exp(\hat{S} - \hat{s}\hat{n})(2 - \hat{s}\hat{n}),$$

$$\phi_3 = \frac{d\lambda}{df_1} = -\hat{n}\exp(\hat{S} - \hat{s}\hat{n})(1 - \hat{s}\hat{n}),$$

$$\phi_4 = \frac{d\lambda}{df_2} = -\hat{n}.$$

We have used the data in figures 2 and 5 of Rodriguez (1989) and nonlinear regression techniques (Gallant, 1975) to estimate the covariance term for the survival model ($Cov(\hat{S}\hat{s})$) and the variances and covariances for the fecundity model ($Var(\hat{f}_1)$, $Var(\hat{f}_2)$, and $Cov(\hat{f}_1\hat{f}_2)$), with the results that $Var(\hat{f}_1) = 6.86 \times 10^{-6}$, $Var(\hat{f}_2) = 3.25 \times 10^{-7}$, $Cov(\hat{S}\hat{s}) = 1.59 \times 10^{-6}$, and $Cov(\hat{f}_1\hat{f}_2) = -1.3 \times 10^{-6}$. The other terms in equation 3.6, which are given in Rodriguez (1989), are $Var(S) = 0.00139$ and $Var(s) = 3.6 \times 10^{-9}$. The parameter estimates were $f_1 = 0.1125$, $f_2 = 0.000855$, $S = -0.5106$, $s = 0.001335$. The estimate of the variance of $\hat{\lambda}$ obtained by substituting these values into equation 3.6 is 0.00808. The confidence interval is derived assuming that the estimated eigenvalue is normally distributed (e.g., 1.96 times the square root of the variance).◆

used to estimate population growth rates from experimental observations.

There are several other studies that use specific models to estimate stability, especially for laboratory populations of blowflies (Stokes et al., 1988) and *Tribolium* (Costantino and Desharnais, 1991; Dennis et al., 1995; Costantino et al., 1997). We discuss these studies in chapters 4 and 5, respectively.

Models Estimated from Data

For many populations, especially natural populations, the most appropriate model of population dynamics is often unknown. In such cases, one can use observations on population size variation over time to estimate the best population dynamic model. The techniques we describe here are all different variants of regression analysis. We describe several techniques for objectively choosing the best model. Once the model has been chosen and its parameters estimated, the stability of the population may be inferred from the same sort of techniques used previously.

Turchin (1990; Turchin and Taylor, 1992; Ellner and Turchin, 1995) has used a technique called the response surface method (RSM). The general form of the model is

$$\log(N_t/N_{t-1}) = P_q(N_{t-1}^{\theta_1}, N_{t-2}^{\theta_2}, \ldots, N_{t-d}^{\theta_d}) + e_{t'} \qquad (3.7)$$

where P_q is a polynomial of degree q, and θ are set to a range of values (e.g., to $-1, -0.5, 0, 0.5, \ldots 3$), e_t are exogenous factors (e.g., weather), and d is the embedding dimension that depends on factors such as age structure and life stage interactions (see chapter 2). Expressed in this way standard linear regression techniques may be used to estimate the parameters of equation 3.7. For instance, when analyzing the *Drosophila* data collected by Rodriguez (1989), the embedding dimension is 1 because there was no age structure, and egg numbers were counted. As a result, the following model was used:

$$\log(N_t/N_{t-1}) = a_0 + a_1 N_{t-1}^{\theta}$$

Turchin's estimate of stability was unaffected by making the polynomial second order.

A difficult question that must be answered when applying equation 3.7 is what values d and q should be. In seeking an answer to this question, one should be guided by the general rules of variable selection in regression analysis. If one looks simply at the proportion of the total variance explained by the regression model, R^2, this quantity typically increases with increasing d and q. In the limit, one can derive an nth-order polynomial that passes through all $n + 1$ points in the regression. This interpolating polynomial typically gives poor estimates of future observations. As with many estimation problems, selection of the "best" regression model involves weighing trade-offs between variance and bias. In box C we review several techniques that have been proposed to aid in evaluating different regression models. These techniques all attempt to achieve a balance between complexity and increased congruity between observations and predictions.

We next apply the response surface method to the *Drosophila* population data in figure 3.2. We have fit these data to first-, second-, and third-order models. All models include a single constant term. In this method, two regression parameters for each embedding dimension were estimated for the independent variables N_t^θ and $N_t^{2\theta}$. Separate models were examined with θ set to -1.5, -1.0, -0.5, 0, 0.5, 1.0, 1.5, 2.0, 2.5, and 3.0. For each model, we estimated V_2, C_p, and *PRESS* to aid in evaluating the best model. *PRESS* was computed by numerically removing one observation at a time and estimating the regression coefficients on the remaining data. This must be done carefully, since the deletion of a single observation in the time series often affects several results. When fitting a third-order model, for instance, when N_t is deleted one can no longer predict N_{t+1}, N_{t+2} and N_{t+3}. The results are shown in table 3.1. (Because the models with $\theta > 1.5$ uniformly performed poorly, these results are not shown.)

The performances of V_2 and C_p are very similar with these data because they both depend on the residual sum

---

BOX C

VARIABLE SELECTION IN REGRESSION MODELS:

For this discussion, we change our notation so the results can be presented in a more general setting. Suppose we have n observations of a dependent variable, y_i ($i = 1, \ldots, n$), that are linear functions of k-independent variables $x_1, x_2, \ldots, x_k$ and have a common variance of σ_2. For population growth models, y_i might be the population size at some time, t, and the x_i's might be previous population sizes or populations sizes squared. The following is a model for the ith observation:

$$Y_i = \beta_1 x_1 + \beta_2 x_2 + \cdots + \beta_k x_k$$

For the entire data set, the model may be written in matrix notation as

$$\mathbf{Y} = \mathbf{X}\beta,$$

where $\mathbf{Y}$ is an $n \times 1$ vector of observations, $\mathbf{X}$ is the $n \times k$ design matrix, and β is the $k \times 1$ parameter vector that we estimate by least-squares techniques. The least-squares estimate of β is given by

$$\hat{\beta} = (\mathbf{X}^T \mathbf{X})^{-1} \mathbf{X}^T \mathbf{Y},$$

where T denotes a matrix transpose. The covariance of $\hat{\beta}$ is given by

$$Cov(\hat{\beta}) = (\mathbf{X}^T \mathbf{X})^{-1} \sigma^2.$$

Suppose we assume that the last $k - p$ parameters are zero (e.g., $\beta_{p+1} = \ldots = \beta_k = 0$). Under this assumption, let $\tilde{\beta}$ be the least-squares estimate of β. It turns out that the variance of predictions based on the more complicated model is greater than the variance of predictions based on the simpler model. In other words, the model predictions using $\hat{\beta}$ have greater variance than predictions based on $\tilde{\beta}$,

$$Var(\lambda^T \hat{\beta}) \geq Var(\lambda^T \tilde{\beta}),$$

BOX C cont.

where λ is a vector of the independent variables (Walls and Weeks, 1969). Of course if the last $k - p$ parameters are not zero, then $\tilde{\beta}$ will be biased. Hence, because the reduction in bias may not compensate for the increased prediction variance, the practical problem is deciding when to stop adding parameters to the model.

One way to incorporate the joint properties of bias and variance of an estimator is through the mean-squared error that equals the variance plus the bias squared. One method for variable selection proposed by Mallows (1973) is to choose parameters that minimize the mean-squared error of the predictor variables $\hat{Y}_i$. Let the residual sum of squares (RSS) for the p-parameter model be $\sum(Y_i - \hat{Y}_i)^2$, then Mallows c_p is defined as $RSS_p/\hat{\sigma}^2 - [n - 2(p + 1)]$. The model with the smallest value of c_p is selected. The variance, $\hat{\sigma}^2$, is estimated from the full model with all k parameters as $RSS_k/(n - k)$.

The statistic called prediction sum of squares or PRESS is based on the ability of the regression model to accurately predict new observations (Allen, 1974). PRESS is a form of cross validation computed by deleting the ith observation from the data set and then using the remaining $n - 1$ data points to estimate, $\hat{\beta}_{-i}$. PRESS is then computed as

$$\frac{1}{n} \sum_i \left[Y_i - \hat{Y}_{-i}\right]^2,$$

where $\hat{Y}_{-i}$ is the prediction based on the estimates, $\hat{\beta}_{-i}$.

Ellner and Turchin (1995) proposed another cross validation statistic:

$$V_c = \left\{ \frac{RSS}{n - pc} \right\}^2$$

In their study, they set $c = 2$. ◆

TABLE 3.1. Results of the response surface method on the two populations of *Drosophila*[*]

Model/θ	Population 8			Population 14		
	V_2	C_p	*PRESS*	V_2	C_p	*PRESS*
First Order						
−1.5	0.036	74	3.0	0.017	69	1.4
−1.0	0.033	69	0.86	0.015	62	0.37
−0.5	**0.015**	**36**	0.79	**0.0069**	**34**	0.31
0	**0.015**	37	0.35	0.0089	42	0.17
0.5	0.017	41	0.20	0.0091	43	**0.13**
1.0	0.035	73	**0.17**	0.024	88	0.14
1.5	0.049	92	0.19	0.032	105	0.15
Second Order						
−1.5	0.031	58	4.9	0.0057	24	1.7
−1.0	0.024	48	1.1	0.0048	20	0.41
−0.5	**0.0064**	**13**	4.0	**0.0023**	**7.0**	0.11
0	0.0079	17	0.72	0.0024	7.3	**0.078**
0.5	0.0071	15	**0.15**	0.0025	7.8	0.089
1.0	0.039	67	0.18	0.033	90	0.18
1.5	0.062	92	0.19	0.040	101	0.15
Third Order						
−1.5	0.0068	12	0.41	0.0051	17	1.1
−1.0	0.0074	13	0.20	0.0053	18	0.24
−0.5	**0.0056**	**9.0**	1.3	**0.0036**	**12**	0.81
0	0.0059	9.8	0.28	0.0037	**12**	0.13
0.5	0.0066	11	0.16	0.0047	16	0.11
1.0	0.024	39	**0.14**	0.0063	21	0.098
1.5	0.048	63	**0.14**	0.0085	28	**0.064**

[*] The minimum value for each statistic is shown in bold.

of squares. On the other hand, for these populations *PRESS* suggests that the best model differs from the best model identified by V_2 and C_p. We determined the largest eigenvalue for the range of models that received support by any of the selection statistics (table 3.2).

TABLE 3.2. The largest eigenvalue (or modulus in the case of complex numbers) for several response surface method models and two populations of *Drosophila*

Model/θ	Population 8	Population 14
Second Order		
−0.5	—	0.77
0	—	0.75
Third Order		
−0.5	0.74	—
0	−0.60	0.75
0.5	−0.63	0.69
1.0	0.69*	0.63

* Complex eigenvalue
— not one of the best models

In all cases, the different models predicted a stable point equilibrium. In this regard, the results in table 3.2 are also consistent with the stability estimates from the single-generation experiments in box A. In the case of population 14, the numerical estimate of the largest eigenvalue was also consistent between models. For population 8, however, the largest eigenvalue was positive, negative, or complex, depending on the model used. Obviously for these regression techniques to be useful, it is important that their qualitative predictions of stability do not change radically as the model structure is changed slightly. Little research has addressed this particular problem, and a more systematic exploration of the RSM techniques ought to be pursued.

TIME-SERIES ANALYSIS

Even populations that are governed by simple models of density-dependent growth vary over time due to random phenomena. If the expected value of the population size is independent of time, then the stochastic process is stationary.

The deterministic component of these types of stochastic processes can be inferred from time-series analysis.

Turchin (1990; Turchin and Taylor, 1992) has been responsible for the most recent use of time series for the elucidation of population stability (for general reviews of applications to population dynamics, see Royama, 1992, and Kendall et al., 1999). As a tool in ecology, time series has been used much earlier to look at the cyclic nature of the predator-prey cycles (Moran, 1953) as well as to model population dynamics in variable environments (Roughgarden, 1975). Turchin and Taylor (1992) described several general patterns for the auto-correlation function (box D) that are expected under different types of population regulation models.

In chapter 2, the departures from an equilibrium, ε_t, for the discrete-time models were represented as a first-order autoregressive process,

$$\varepsilon_{t+1} = a\varepsilon_t,$$

where a is the first derivative of the density-regulating function evaluated at the equilibrium point. If we assume that the mean of ε_t is zero and the variance is σ^2, then we get the following:

$$\rho_\varepsilon(k) = \frac{Cov(\varepsilon_t \varepsilon_{t-k})}{\sigma^2} = \frac{E(\varepsilon_t \varepsilon_{t-k})}{\sigma^2} = \frac{E(a^k \varepsilon_{t-k} \varepsilon_{t-k})}{\sigma^2}$$
$$= \frac{a^k \sigma^2}{\sigma^2} = a^k$$

Populations with positive eigenvalues ($0 < a < 1$) produce positive autocorrelations that decline geometrically as observations become more distant (fig. 3.3). This result implies that observations that are closest tend to be similar in value, whereas more distant observations show a weaker resemblance. With negative eigenvalues ($-1 < a < 0$) the sign of the correlation changes with each lag. Thus, the correlation is negative with odd lags and positive with even lags, and all correlations decline in magnitude with increasing lags (see

91

BOX D

TIME SERIES, THE AUTOCORRELATION
AND SPECTRAL DENSITY FUNCTIONS:

A statistical time series may be a continuous or discrete time varying function, $x(t)$, that is subject to random variation (Jenkins and Watts, 1968). Observations made at different times are generally not independent of each other but may be related by some linear or nonlinear function. When the underlying process that controls the time series reaches an equilibrium or steady state, the process is stationary. The value of $x(t)$ does not depend on the absolute time for a stationary process. Most of the techniques and analyses of time series require the assumption of stationarity. An informative property of a time series is the autocorrelation function, $\rho_x(k)$, which defines the correlation between observations separated by k time units (k is often referred to as the lag). For a stationary process, the autocorrelation function depends only on the time separating the observations, not the absolute time. We define the autocorrelation function as

$$\rho_x(k) = \frac{Cov\big(x(t), x(t-k)\big)}{\sigma_{x(t)}\sigma_{x(t-k)}},$$

where $\sigma_{x(t)}$ is the standard deviation of the random variable at time t. The stationarity assumption leads to the natural conclusion that $\sigma_{x(t)} = \sigma_{x(t-k)} = \sigma$ and, therefore:

$$\rho_x(k) = \frac{Cov\big(x(t), x(t-k)\big)}{\sigma_2}$$

For finite data sets, the number of observations available to estimate the autocorrelations decreases with increasing k. Suppose population sizes are estimated for 10 generations yielding $N_1, N_2, \ldots N_{10}$. From these observations, there are only two pairs of points to estimate $\rho(8)$ (N_1, N_9 and N_2, N_{10}) but nine pairs to estimate $\rho(1)$. For this reason, the most accurate estimates of autocorrelations are those at the small lags.

BOX D cont.

The autocorrelation is one technique for studying time series and is usually referred to as an analysis of the time domain. The variance of a time series can be decomposed into sine and cosine waves of different frequencies. These techniques are called analyses in the frequency domain. The spectral density function describes the relative contribution to the total variance of periodic functions with different frequencies. Random variables that are uncorrelated over time have roughly equal contributions to their variance from periodic functions of all frequencies. In this sense, these random variables resemble white light, which results from mixing light of many wavelengths. Consequently, these types of random variables are sometimes called white noise. Some time series may have strong periodic components, and these will show up as a peak in the spectral density function. For time series with observations made at M regular time intervals, the highest frequencies that can be detected are signals with periods of two time units or frequencies of 0.5 cycles per time unit. Higher frequency oscillations will simply be undetectable because the signal may undergo several unobservable cycles between sample points. The lowest frequencies will be oscillations with periods equal to M or frequencies of M^{-1} cycles per time interval. Again, lower frequency oscillations cannot be detected since the sample size is not sufficient to observe at least one complete cycle. The techniques for estimating spectral density functions are somewhat complicated (see Jenkins and Watts, 1968, chapter 6, for details), and they typically use smoothing functions that attempt to balance the problems of bias and variance. These smoothing functions, or windows, can reduce the variance of the estimator considerably by using many adjacent estimates of spectral density, although this process also introduces bias.◆

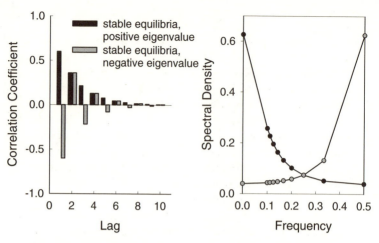

F IG . 3.3. Autocorrelation and spectral density for a first-order autoregressive process. For a stable population with positive eigenvalues, the autocorrelations are always positive and decrease with increasing lag. For stable populations with negative eigenvalues, the autocorrelations are negative for odd lags, are positive for even lags, and decrease in magnitude with increasing lag. Populations with positive eigenvalues exhibit most of their variation at low frequencies, whereas populations with negative eigenvalues are dominated by high-frequency variation.

fig. 3.3). This type of behavior is generated by the oscillatory approach to the equilibrium.

With a little more work, we can also derive the expression for the spectral density function for the first-order autoregressive process (Jenkins and Watts, 1968, p. 228) as

$$\Gamma_\varepsilon(f) = \frac{\sigma^2}{1 + a^2 - 2a\cos(2\pi f)},$$

where σ^2 is the variance of the random noise, and f is the frequency. For positive values of a (the stability determining eigenvalue), the spectrum is dominated by low-frequency signals (fig. 3.3), whereas for negative values of a, the spectrum is dominated by high-frequency components.

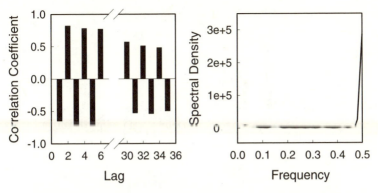

FIG. 3.4. Autocorrelation and spectral density estimated from the adult data in figure 2.9. These data were in a two-point cycle with a relatively small amount of noise added to each generation's population size. Population sizes separated by an even number of generations (lags) are close to the same equilibrium point (either the valley or the peak) and thus show strong positive correlations. Population sizes separated by an odd number of generations are at opposite positions (one at a peak, the other at the valley) and thus show a strong negative correlation. The spectral density is dominated by the high-frequency oscillations.

We next consider populations in a stable cycle. Suppose the population is at a stable two-point cycle with equilibrium points $\hat{N}_1, \hat{N}_2$. Population sizes separated by an even number of time units will be close to the same equilibrium point and thus positively correlated with each other, whereas the population sizes separated by odd numbers of time units will be at the alternate equilibria and thus negatively correlated. The magnitude of these correlations should also get weaker with increasing lag due to multiple time intervals of intervening random noise. To illustrate this, figure 3.4 shows the autocorrelation function for the adult data in figure 2.9. In this example, the relatively small amount of environmental noise results in very slow decay of the magnitude of the correlation function with time. If there were no environmental noise, the correlation would be 1.0 for all even lags and −1.0 for all odd lags.

95

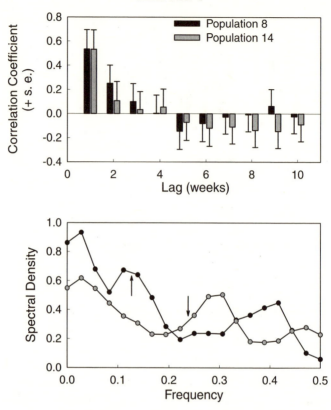

FIG. 3.5. Autocorrelation coefficient for the two populations of *Drosophila* shown in figure 3.2. The data was log transformed, and then any linear trend was removed. In both populations, a strong positive correlation between neighboring observations rapidly decays to zero as the observations become more distant, but then become slightly negative. The spectral density function for population 8 shows a peak around a frequency of 0.12 to 0.14. Population 14 shows a peak around 0.28 to 0.30. The arrows show the frequency of expected peaks from the stability analysis (box A).

These techniques are now applied to the *Drosophila* data analyzed previously. Results shown in figure 3.5 demonstrate that neither population appears to be in a two-point cycle nor in a simple oscillatory approach to equilibrium. In fact, the patterns are most similar to the stable equilibrium with a positive eigenvalue except for the peaks in the spectral density function, which suggest middle-range oscillations. These periodicities may in fact be due to the complex eigenvalues that create an oscillatory approach to equilibrium as outlined in box A. The arrows show the predicted frequencies of these oscillations that are reasonably close to the observed peaks. The difference between the observed and predicted frequency peaks is greater for population 14 than for population 8. This may be due to the use of densities that did not bracket the carrying capacity and hence provided less accurate information about the local linear dynamics of population 14.

CHAOS

The great attraction of chaos for population biologists is that data from real populations often look more similar, at least superficially, to chaotic trajectories than to the trajectories predicted by our simple models. There is little argument over the presence of noise—from environmental factors and in estimates of population size—in natural populations. However, it is important to know if most variation in population numbers is due to these extrinsic sources of noise, or if the noise is generated internally by the density-regulating mechanisms. It is important to keep in mind that the dynamics of all real populations are stochastic. For that reason, the dynamics of real populations cannot be chaotic since chaos is a property of deterministic systems. Stochastic systems may have properties in common with chaotic systems, such as positive Lyapunov exponents, but it is important to distinguish the behavior of the stochastic system from that of the underlying deterministic system.

Time Series

Recently, the patterns of spectral density functions have been used to assess the likelihood of chaotic dynamics

97

(Cohen, 1995). Because most long-term records are for natural populations, an important component to this evaluation is some impression of the spectral density functions due to random environmental variation. At first it might seem that most environmental noise would be white. This would clearly depend to some extent on the sampling period. Thus, local temperature will show strong correlations from one time point to the next when sampled at 24-hour intervals. Over yearly intervals, however, the correlation will be weak or zero. Steele (1985) presents data suggesting that atmospheric temperature shows white-noise variation for intervals of time up to about 50 years. With longer intervals, there is an increasing contribution of periodic components with long wavelengths or low frequencies. This increase in the spectral density at low frequencies is sometimes referred to as a red spectrum, following the analogy to frequencies of visible light. This pattern is even more pronounced in ocean temperatures that show a spectral density that continuously increases in proportion to $1/\text{frequency}^2$.

If these very long-term environmental fluctuations force ecological systems to jump between alternative equilibria, then the long-term spectrum of population numbers will also exhibit a red spectrum (Steele, 1985). The direct analysis of long time series of Chinese locust (Sugihara, 1995) and indirect inferences from several other species (Pimm and Redfearn, 1988; Ariño and Pimm, 1995) suggest that there may be significant redness in these spectra. Cohen (1995) presented a new ecological dilemma when he noted that many simple population-growth models yield significantly blue spectra under conditions that produce chaotic dynamics. As we saw earlier in this chapter, populations exhibiting stable cycles or an oscillatory approach to equilibrium also exhibit blue spectra (assuming small levels of white noise). On the other hand, populations with a strong stable equilibrium point exhibit red spectra (again assuming low levels of environmental white noise).

What should be made of Cohen's observations? Sugihara (1995) suggests that they raise the specter of environmental determination of population patterns (red spectra) versus population regulation (blue spectra). We feel that the significance of these findings has been somewhat overstated and that there is ultimately little utility of time-series spectra for evaluating the potential for chaos in natural populations. Several points need to be considered. The first is the utility of the existing time series. The Chinese locust data were recorded over one of the longest time periods—about 1000 years. While one can make out a slight increase in the magnitude of the spectrum at low frequencies, this increase is not nearly as dramatic as similar data for physical factors such as temperature. One can imagine that these data might be subject to long-term cycles in their quality. Thus, due to political and financial resources the census data may be subject to periods of good collection (in which a large fraction of the population is accounted for) and periods of poor collection (in which a much smaller fraction of the population is accounted for). This type of fluctuating effort may resemble the ecological model developed by Steele (1985) that also gave rise to a red spectrum. Currently it appears that only a small fraction of natural populations can be classified as chaotic (see chapter 7). Other models of population dynamics can produce red and white spectra (White et al., 1996). Thus, the observation of a red spectrum may eliminate a certain class of chaotic models, but it is not strong evidence of the lack of chaos, nor for the primacy of environmental effects.

Detecting Chaos

Schaeffer (1984) examined the long-term data on lynx skins shipped by the Hudson Bay Company in Canada. In this study, he used three-year running averages to construct a third-order model by the technique of cubic splines. From the resulting model, he then inspected the trajectories in three-dimensional figures to see if one could detect evidence for folding and stretching of the trajectories. When trajectories fold and stretch, they give rise to the sensitive dependence on initial conditions, which is a hallmark of chaos. The qualitative manner in which these techniques need to be applied limits their general utility.

Sugihara and May (1990) used a different approach to infer chaos. They noted that for large (i.e., > 500 observations) time series, nonlinear models fit to these data typically did well predicting future observations, at least for a dozen or so time intervals, unless the trajectories were chaotic. The sensitivity of chaotic dynamics to initial conditions means that predictive power will be lost quickly. Thus, in plots of the correlation coefficient between predicted and observed population size versus the number of time intervals in the future, Sugihara and May suggest that steadily declining correlations over about 12 time intervals indicates chaotic dynamics. Sugihara and May conclude that the number of measles cases in New York City and the number of diatoms off the pier in La Jolla are chaotic, whereas chickenpox cases in New York City are not chaotic. The practical limitation of these techniques is the need for large samples.

Another interesting procedure for detecting chaos has been described by Ellner and Turchin (1995). The ultimate goal of this technique was to estimate directly the Lyapunov exponent of the nonlinear dynamic model. Chaotic populations of course possess positive Lyapunov exponents. These techniques appear to work well with data sets of modest size (~50 observations). Ellner and Turchin proposed estimating a population dynamic model from the observations as described earlier in this chapter. They relied on three general models: the response surface model, feedback neural networks, and thin plate splines. Each method uses very general nonlinear equations, and no arguments from first principles favor one method over another. However, the response surface method typically requires fewer parameters and is particularly helpful for small data sets. In simulations, Ellner and Turchin found no substantial differences among the performances of each of these models.

Ellner and Turchin suggested using the V_2 statistic to choose the best model. As we have seen for the *Drosophila* data analyzed in this chapter, however, other criteria, such as PRESS, do not necessarily identify the same model as best. We suggest using a variety of techniques for identifying the best model, and if no single model "wins", then determining the Lyapunov exponent from the range of best models. We think this is a prudent method for several reasons. Obviously, if different objective criteria cannot distinguish among several models, they should all be examined. PRESS and V_2 help identify models that may provide the best future predictions, but that does not necessarily mean that they will provide the most reliable estimates of the stability of the studied populations. For this reason, it is worthwhile to look at several models to ensure that results are consistent. Otherwise, our conclusions about the existence of chaos are only as strong as our belief that the chosen model is the proper one for the study population.

Once a model is chosen, the Lyapunov exponent is estimated in a manner similar to the technique used in chapter 2. We illustrate this with a second-order model. The extension to higher-order models is straightforward. Suppose we have m observations labeled $\tilde{N}_0, \tilde{N}_1, \ldots \tilde{N}_{m-1}$. The model of population growth is

$$N_t = g(N_{t-1}, N_{t-2}). \tag{3.8}$$

Let the vector $\mathbf{N}_t$ be $(N_t, N_{t-1})^T$. Then equation 3.8 may be rewritten as

$$\mathbf{N}_t = \begin{pmatrix} g_1(\mathbf{N}_{t-1}) \\ g_2(\mathbf{N}_{t-1}) \end{pmatrix} = \begin{pmatrix} g(N_{t-1}, N_{t-2}) \\ N_{t-1} \end{pmatrix}.$$

Box E

Vector Norms:

If $\mathbf{v}$ is an n-dimensional vector with real or complex elements, a vector norm will have the following properties: (1) $\|\mathbf{v}\| > 0$, unless $\mathbf{v} = \mathbf{0}$, (2) if c is a scalar, then $\|c\mathbf{v}\| = |c|\|\mathbf{v}\|$, (3) for any vectors $\mathbf{v}$ and $\mathbf{u}$, $\|\mathbf{v} + \mathbf{u}\| \leq \|\mathbf{v}\| + \|\mathbf{y}\|$. Condition (3) is also known as the triangle inequality. The Euclidean norm (or distance) is defined as

$$\|\mathbf{v}\|_2 = \left(\left|\nu_1\right|^2 + \cdots + \left|\nu_n\right|^2\right)^{1/2}.$$

Other commonly used norms include

$$\|\mathbf{v}\|_1 = \sum_{i=1}^{n} |\nu_i| \text{ and } \|\mathbf{v}\|_\infty = \max_i |\nu_i|. \; \blacklozenge$$

We can then define the Jacobian matrix as

$$\hat{\mathbf{J}}_{t-1} = \begin{bmatrix} \dfrac{dg_1(\mathbf{N}_{t-1})}{dN_{t-1}}\bigg|_{\bar{\mathbf{N}}_{t-1}} & \dfrac{dg_1(\mathbf{N}_{t-1})}{dN_{t-2}}\bigg|_{\bar{\mathbf{N}}_{t-1}} \\[2ex] \dfrac{dg_2(\mathbf{N}_{t-1})}{dN_{t-1}}\bigg|_{\bar{\mathbf{N}}_{t-1}} & \dfrac{dg_2(\mathbf{N}_{t-1})}{dN_{t-2}}\bigg|_{\bar{\mathbf{N}}_{t-1}} \end{bmatrix}.$$

The estimated Lyapunov exponent is

$$\hat{\lambda} = \frac{1}{m-1} \log \left\|\hat{\mathbf{J}}_{\mathbf{m-1}}\hat{\mathbf{J}}_{\mathbf{m-2}}\cdots\hat{\mathbf{J}}_{\mathbf{1}}\mathbf{v}\right\|, \tag{3.9}$$

where $\mathbf{v}^{\mathrm{T}} = (1, 0)$. The double bars indicate any vector norm (box E). In simulated trajectories of 100 observations, these techniques were effective at distinguishing chaotic from non-chaotic dynamics. However, these simulations were done on a limited set of models, and more work on these techniques is needed.

It is worth contrasting equation 3.9 to the technique used in figure 2.4 to estimate the Lyapunov exponent. In figure 2.4, the deterministic model was iterated many times to determine if nearby trajectories tend to diverge from each other. In equation 3.9, this behavior is estimated along the actual orbit of the observed population sizes. These observations clearly consist of both the deterministic portion of population dynamics and the random component. In fact, we might view the observed population sizes as a realization of the stationary distribution of population size. The method developed by Ellner and Turchin (1995) is therefore a stochastic estimate of stability (see chapter 1). Another technique for estimating stochastic Lyapunov exponents has recently been described by Dennis et al. (in press). Their technique iterates a stochastic model of population dynamics and is similar to the technique used in figure 2.4. The technique proposed by Dennis et al. differs from that of Turchin and Ellner in three ways: (1) Dennis et al. use a mechanistic-based model fit to data, whereas Ellner and Turchin use nonparametric models fit to data. Dennis et al. applied their technique to experimental *Tribolium* data for which there is a good theoretical understanding of the growth model (see chapter 5). In contrast, Turchin and Ellner apply their techniques to many natural populations for which this level of understanding is seldom found. (2) Dennis et al. iterate a stochastic model and evaluate Jacobian products until convergence, whereas Turchin and Ellner evaluate the Jacobians using the observed data trajectory. This requires Dennis et al. to make some assumption about the form of the random noise, whereas Ellner and Turchin are using the observations as an empirical estimate of that distribution. (The consequences of assuming the wrong distribution have not yet been studied.) (3) Dennis et al. demonstrate a bootstrapping technique to provide confidence intervals for the Lyapunov exponents and stochastic Lyapunov exponents.

Dennis et al. (in press) prefer to estimate separately the deterministic and stochastic parts of a population's dynamics. They reason that there is inherent interest in determining the extent to which a population's overall behavior is a consequence of the underlying biology that determines the nonlinear growth equations. In the experimental *Tribolium* systems studied by Dennis et al., for instance, they estimate that the deterministic component of population dynamics explains 93 to 99 percent of the observed variability depending on the life stage examined. In all likelihood, natural populations would have a larger contribution from random forces.

On the other hand, Dennis et al. note the potential for stochastic noise to affect the final dynamics of populations. For instance noise may cause populations to spend long periods of time near unstable equilibria or, in cases where there are multiple domains of attraction, to bounce between these alternative states.

In many ways, the final description of population dynamics involves the consideration of elements analogous to those that appear in descriptions of the evolution of populations: natural selection and genetic drift. Evolution depends on both the deterministic force of natural selection and the random force of genetic drift. We find situations in which one or the other force is likely to dominate evolution: drift dominates in small populations, while in large populations selection controls the fate of novel beneficial traits (e.g., antibiotic resistance in bacterial populations). Important synthetic theories of evolution emphasize the joint role of both forces. Wright's shifting balance theory of evolution uses the potential of drift to place populations near different domains of attraction and thus to "permit" evolution to explore the adaptive landscape.

We think that just as evolutionary biology finds it useful to distinguish stochastic and deterministic forces, population dynamics will benefit from evaluating these components separately. There are several reasons for adopting this approach: (1) In most experimental systems, it is the deterministic aspects of population dynamics that have been manipulated, although in the future we may find experimental work that attempts to manipulate random aspects of the

TABLE 3.3. Stochastic Lyapunov exponents for the *Drosophila* populations 8 and 14 shown in figure 3.2*

θ	Population	Stochastic Lyapunov Exponent
−0.5	8	−0.49
	14	−0.13
0	8	−0.09
	14	−1.3
0.5	8	−0.52
	14	−0.27

* The second-order models used in table 3.1 were used with several different values of θ.

environment. (2) Evolution of life histories will affect the deterministic aspects of population dynamics. (3) Detailed study of environmental variation, at least in natural populations, is likely to be relevant only to specific geographic regions and only for specific periods of time. As a result, an understanding of this variation will provide less general knowledge than we can derive through an equivalent study of the deterministic aspects of population dynamics.

We already know from our analysis of the deterministic models that have been fit to the *Drosophila* data (fig. 3.2) that the equilibrium points are stable (table 3.2). We may now use the method of Ellner and Turchin (1995) to determine if the stochastic Lyapunov exponent is positive (table 3.3). All stochastic Lyapunov exponents shown are negative. Thus, even with environmental noise, proximate trajectories will remain close to each other.

Time-series data from natural populations are sometimes collected at monthly intervals rather than yearly. Samples collected in this way may reflect seasonal variation in addition to other sources of variation. Ellner and Turchin (1995) suggest that forcing the population dynamic model to explain this regular source of variation can lead to spurious inferences of chaotic dynamics. They suggest adding to the regression models the periodic parameters $\cos(2\pi t/12)$ and

$\sin(2\pi t/12)$. Using this technique, they found that monthly records of measles were weakly stable with the inclusion of the periodic function but chaotic without it. At present, the relative importance of chaos in the dynamics of natural populations under debate. We return to this issue in our discussion of natural populations in chapter 7.

APPENDIX TO CHAPTER THREE

TABLE 3.1A. Variation in population size in the two populations of *Drosophila melanogaster* graphed in figure 3.2

Week	Population 8	Population 14
1	100	100
2	92	100
3	364	123
4	727	729
5	422	1201
6	1433	891
7	1082	968
8	1093	1553
9	898	1261
10	719	1336
11	833	1391
12	870	1330
13	997	1157
14	631	1004
15	1405	1393
16	906	838
17	961	1114
18	1066	789
19	895	986
20	801	1074
21	923	1302
22	1002	1309
23	963	1170
24	1135	855
25	1186	1065
26	800	794
27	808	1030
28	845	912
29	741	845
30	867	659
31	504	861
32	820	553
33	428	853

Table 3.1A. Cont.

Week	Population 8	Population 14
34	862	966
35	974	754
36	1058	902
37	841	1062
38	762	1108

TABLE 3.2A. Observed survivors ($g_1(N^*)$) and emerging progeny ($g_2(N^*)$, $g_3(N^*)$, $g_4(N^*)$) from the single-density experiments for populations 8 and 14 of *D. melanogaster*

Population	Density (N^*)	Replicate	$g_1(N^*)$	$g_2(N^*)$	$g_3(N^*)$	$g_4(N^*)$
Line 8	750	1	361	230	274	173
		2	392	207	311	104
		3	321	334	326	81
		4	322	366	290	84
		5	103	266	229	73
		6	89	309	227	47
	1000	1	191	436	298	63
		2	108	315	238	62
		3	138	304	201	94
Line 14	750	1	45	353	314	289
		2	39	333	298	269
		3	295	258	373	147
		4	304	222	349	142
		5	109	287	460	378
		6	91	286	404	355
	1000	1	308	197	447	260
		2	139	160	322	408
		3	167	206	332	367

CHAPTER FOUR

Blowflies

In the 1950s, the Australian entomologist A.J. Nicholson conducted a series of experiments aimed at studying the dynamics of laboratory populations of the Australian sheep blowfly *Lucilia cuprina* Wied. under various types of food regimes and demographic and environmental perturbations (Nicholson, 1954a, 1954b, 1957). These experiments were conducted during the heyday of the debate, to which we alluded in chapter 1, about the importance of density-dependent versus density-independent mechanisms in population regulation. Nicholson attempted to demonstrate experimentally that populations were self-regulating and could compensate for various perturbations to their numbers. However, in light of the debate just mentioned, some of Nicholson's concerns today seem a little anachronistic. Ironically, the regular cycles in population numbers that Nicholson observed, and which have been a continuing focus of attention and interest among population ecologists, were to him "only of secondary importance." Nevertheless, the data collected by Nicholson in the course of his experiments have become well known as a textbook example of how populations can fluctuate violently in numbers even in constant environments. They have also motivated several modeling efforts in more recent times (May, 1973; Brillinger et al., 1980; Gurney et al., 1980; Nisbet and Gurney, 1982; Stokes et al., 1988; Manly, 1990; Gutierrez, 1996). In this chapter we review some of Nicholson's experiments, concentrating on those results that shed some light on how various density-dependent regulatory mechanisms can affect the stability of populations.

110

LIFE HISTORY OF *LUCILIA CUPRINA*
IN THE LABORATORY

In his experiments, Nicholson reared large populations of blowflies at 25°C in perspex cages that could support adult populations of more than 10,000 flies. Under these conditions, eggs hatched in 12 to 24 hours, and the larval stage extended from 5 to 10 days, with the duration depending in part on larval density. The pupal stage took 6 to 8 days. Young adults of *L. cuprina* are not sexually mature until about four days of adult age, and females may be about eight days old before they are able to lay any appreciable numbers of eggs. The adults could live up to 35 days or so. Thus, the total development time, from egg to egg, in these populations was about 20 to 22 days.

In these experiments, laboratory populations of *L. cuprina* were subjected to two major types of food regimes. In one food regime (henceforth referred to as HL: high food levels for larvae, low food levels for adults), larvae were given food well in excess of their requirements, and adults did not have access to this food. The adults, on the other hand, were supplied sugar and water to excess but had their protein supply limited by adding only 0.5 g of ground liver each day to the culture. In *L. cuprina*, adult mortality levels and age-dependent patterns are similar whether they are given sugar, water, and liver or just sugar and water. Life span, however, can be shortened considerably by reducing the amount of sugar supplied to adults. The requirement of adequate protein for adults is important for female fecundity. If the protein intake of females is below a threshold level, they lay no eggs, and fecundity in general declines with decreasing protein intake. In the other main food regime (henceforth referred to as LH: low food levels for larvae, high food levels for adults), adults were given sugar, water, and liver well in excess of their requirements, and larvae did not have access to this food. The larvae in this food regime were provided

111

only 50 g of food per day (25 g per day in some treatments), which would give rise to severe scramble type of larval competition for food.

It is clear that major density-dependent effects in these populations were mediated through food availability for larvae or adults, according to food regime. Both larval and adult life stages in *L. cuprina* were subject to density-dependent mortality. In larvae, there was negligible mortality when excess food was provided; the density-dependence of mortality was, thus, primarily due to density in terms of larvae per unit food. At densities of 5 to 10 larvae per gram of food, 80 to 90 percent of the larvae survived to eclose as adults. Survivorship until eclosion declined rapidly as larval density increased from 10 to 40 larvae per gram and was less than 2 percent for densities greater than about

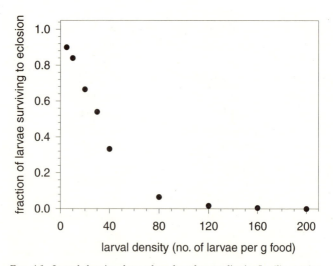

FIG. 4.1. Larval density–dependent larval mortality in *Lucilia cuprina*. Data points are means from several independent replicate vials at each density. Survival until eclosion is also determined in part by pupal mortality, which is ~0.02 and is independent of larval density (data from Nicholson, 1954b).

100 larvae per gram (fig. 4.1). Survival of larvae until eclosion, of course, depends on both larval and pupal mortality,
but pupal mortality in *L. cuprina* was not dependent on larval
density, and in Nicholson's experiments it fluctuated errati-
cally about a mean level of approximately 8 percent. Larval
density also had an effect on the size of eclosing adults,
with substantial reduction in adult size being observed even
at larval densities at which survivorship was not affected so
markedly. Adult mortality in these populations appeared to
increase almost linearly with density, at least over the range
of densities observed in the course of Nicholson's experiments (fig. 4.2). The mortality values in figure 4.2 are for
the fraction of adults dying over a two-day period, and it can

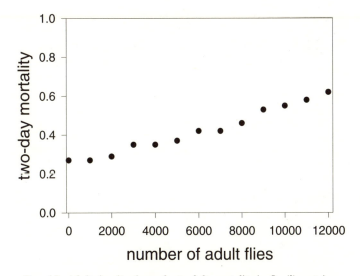

FIG. 4.2. Adult density–dependent adult mortality in *Lucilia cuprina*.
Data points are means averaged over 10 to 20 separate observations of
the number of deaths during two-day periods as a fraction of the number of adults present at the beginning of those two days. The figure is
schematic and depicts the average mortality over a density range. For
example, mortality over two days at densities of 1500 to 2500 adults
would be about 30% (modified after figure 8.2 in Manly, 1990).

113

be seen that at densities above 6000 adults or so, the adult numbers within a cohort would fall quite drastically in just a few days, lowering the mean life span substantially.

Female fecundity in these populations was also adversely affected by increased adult density, and this effect was heightened in the HL food regime, in which the supply of protein to adults was limited (fig. 4.3). The pattern depicted in figure 4.3, which shows the sensitivity of fecundity to adult density in the LH and HL food regimes, is for illustrative purposes only and should be interpreted qualitatively. The data to which the hyperbolic model of fecundity as a function of density have been fitted are average daily fecundity and average adult population size data from a number of treatments within both LH and HL regimes, in which

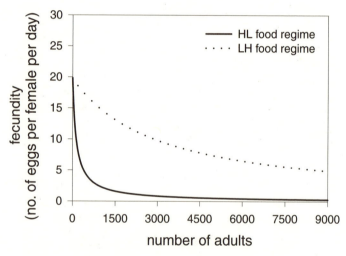

FIG. 4.3. Idealized depiction of the sensitivity of female fecundity to adult density under the LH (unlimited protein for adults) and HL (adult protein supply limited to 0.5 g liver per day) food regimes. The figure shows best-fit curves obtained by fitting a hyperbolic model of fecundity as a function of density ($F(N_t) = a/(1 + bN_t)$) to data on mean numbers of eggs produced daily in populations maintaining different mean densities (data from Nicholson, 1954a).

varying proportions of adults were culled by the experimenters. Thus, large fluctuations in adult numbers, and in fecundity, have been subsumed into the mean values. Our main purpose here is simply to illustrate that, broadly speaking, fecundity declined with adult density in both food regimes, but the sensitivity of this density-dependent response was markedly reduced in the LH food regime. Female fecundity would also have been affected adversely by high larval densities due to a reduction in adult size, which was up to eightfold in these experiments when larval densities were high. The effect of larval density on female fecundity in the LH regime, however, was of far smaller magnitude than the effect due to adult density in the HL food regime (Nicholson, 1957).

Overall, it seems reasonable to conclude that adult density–dependent fecundity and larval density–dependent larval mortality are the main density–dependent regulatory mechanisms in the HL and LH food regimes, respectively (fig. 4.4). In the LH food regime, inhibitory effects of larval and adult densities on female fecundity also probably play a subsidiary role in generating negative feedback. Adult density-dependent adult mortality is likely to be a minor contributor to the density-dependent regulation of population growth in both food regimes.

DYNAMICS OF *LUCILIA CUPRINA* POPULATIONS

In the experiments conducted by Nicholson, populations of *L. cuprina* were established by seeding a cage with 1000 pupae that were essentially allowed to maintain themselves for up to a year or more, with minimal disturbance other than the imposition of an LH or HL food regime. Data were recorded a few weeks after initiating a population to avoid any transient dynamics. The numbers of larvae, pupae, and living and dead adults were recorded daily. Some of the populations within each food regime were also subjected to

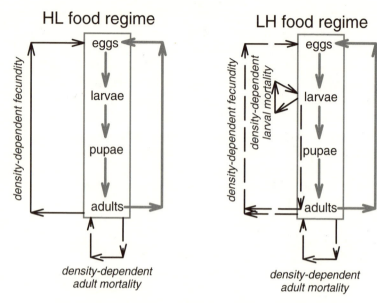

Fig. 4.4. Schematic depiction of the density-dependent regulatory mechanisms acting on populations subjected to LH and HL food regimes. Thick gray arrows represent ontogenetic transformations. Solid and dotted thin black arrows represent relatively strong and relatively weak density-dependent feedback loops, respectively.

demographic perturbation by killing a fixed proportion of eclosing adults each day. Others were subjected to variations on the food regimes, which are discussed later.

From the point of view of stability, the most important result from these experiments was that fairly regular, large-amplitude (3 to 4 orders of magnitude) oscillations in adult numbers, with a periodicity of about 40 days, were seen in both HL and LH food regimes, although the period appeared to be somewhat shorter in the LH food regimes (fig. 4.5). In the HL food regime, minima in adult numbers were as low as a few tens of adults, whereas adult numbers at the maxima routinely exceeded 7000, going as high as 14,000 in some cycles. The magnitude of the maxima was clearly affected by the amount of food provided to the

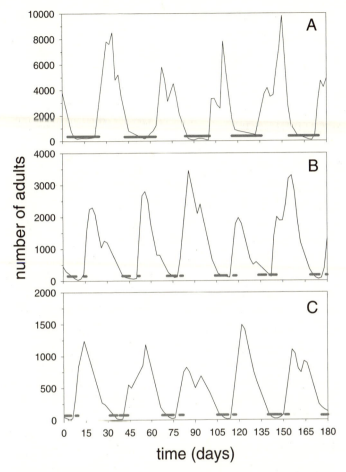

FIG. 4.5. Representative samples of a few cycles of time-series data from three populations, illustrating the regular fluctuations observed both LH and HL food regimes. Data from the initial few weeks of each population would represent transient dynamics and have, therefore, been omitted. Panel A shows data from a population subjected to an HL food regime. Panels B and C show data from populations in LH food regimes with larval food supply held at 50 g and 25 g per day, respectively. Note the different scaling of the y-axis in each panel. Horizontal gray bars represent times that appreciable numbers of eggs were laid (panel A), or egg-laying periods for which the eggs gave rise to an appreciable number of eclosing adults (panels B and C) (modified after Nicholson, 1954b, 1957).

larvae. In an LH food regime in which larvae were limited to 50 grams of food per day, observed maxima were in the range of 2000 to 3500 adults; when larval food supply was held at 25 g per day, however, the range of observed maxima was only 700 to 1500 adults. Although the amplitude of the fluctuations was much reduced in the LH treatments characterized by limited larval food supply, the degree of instability of populations in the two food regimes was not different, with coefficients of variation of adult numbers being approximately 1.0 in all three populations shown in figure 4.5.

In both LH and HL food regimes, appreciable recruitment into the adult stage resulted only from eggs laid during troughs in adult numbers (see fig. 4.5). In the HL food regime, in which protein supply to adults was limited, appreciable numbers of eggs were laid only when adult numbers were very low, enabling at least some females to obtain enough protein for sustaining egg laying. Larval mortality in this food regime was negligible, and pupal mortality was not density dependent. Consequently, the numbers of breeding adults recruited on any day was directly proportional to the number of eggs laid about 20 days before. In the LH food regimes, adults were not limited by protein supply and were consequently relatively free from adult density–dependent control on fecundity (see fig. 4.3). As a result, large numbers of eggs were laid each day, especially during periods of high adult density. However, at the kinds of egg densities reached in the LH cultures ($\sim$175 larvae per gram, on average), larval mortality would be very excessive ($\sim$0.99; see fig. 4.1) except when relatively few adults were laying eggs. Thus, even in the LH food regime, in which adults were not competing for protein sources, significant recruitment into the pool of breeding adults would occur only from a cohort of eggs laid when adult density was extremely low. In addition to the large and regular fluctuations in adult numbers seen in these populations, there is the intriguing observation that the distribution of eggs laid during the period of low density—and which result in significant adult recruitment—is bimodal in the LH food regime (broken grey bars in panels B and C of fig. 4.5), whereas it is unimodal in the HL food regime (solid grey bars in panel A of fig. 4.5). In fact, two distinct peaks in pupal numbers can be seen per cycle in the LH food regime, but they appear at somewhat irregular intervals (data not shown). This is a point that Nicholson did not address, although it has been examined by later workers (Gurney et al., 1980). We discuss this trend in the next section, after discussing here the likely causes of the gross features of the observed dynamics in the HL and LH food regimes.

As noted by Nicholson (1957), it appears that the fluctuations in adult numbers in these populations were caused by a combination of (1) relatively high potential fecundity; (2) extremely strong adult density–dependent recruitment into the pool of breeding adults, albeit by different mechanisms in LH and HL food regimes; and (3) a time delay of about 20 days before the impact of adult density at any given time was felt on recruitment into the pool of breeding adults. The destabilizing effect of this time delay was further exacerbated by the relatively high daily adult mortality rate (see fig. 4.2), resulting in a fairly rapid turnover of the cohorts making up the adult population. Thus, in the HL food regime, large numbers of eggs were laid when adult numbers were very low because relatively more females were able to get enough protein for egg laying. As a result of negligible pre-adult competition, these large numbers of eggs eventually became large numbers of adults after about 20 days. At that point, even though egg production dropped drastically due to increased competition among females for scarce protein supplies, recruitment into the adult life stage continued for several days before the effect of the adult density–dependent reduction in fecundity translated into a sharp drop in adult numbers. Every 20 days or so, adult numbers alternated between very high and very low levels, giving rise to a population cycle of about 40 days periodicity.

119

An essentially similar mechanism was also at work in the LH food regime. Very large numbers of eggs were laid when adult numbers were reasonably high because *L. cuprina* females are quite fecund, and there was no limitation on adult access to protein. The large numbers of newly hatched larvae then underwent severe competition for limited food, with the result that all or most larvae were not able to attain the critical minimum size for successful pupation. Consequently, there was little or no recruitment into the adult life stage. Adult numbers, at the same time, were declining due to mortality, and soon the adult density fell to a point at which the number of eggs laid was low enough to significantly lessen the severity of larval competition. At this point, recruitment into the adult life stage began to increase, and adult numbers again increased to a point at which severe larval competition due to an excess of eggs being laid caused a decline in recruitment into the adult stage. This dynamic thus gave rise to cycles in adult numbers similar to those seen in the HL populations.

Based on this explanation of how LH and HL food regimes give rise to cycles in adult numbers, it is clear that an LL regime in which both larvae and adults are given limited food supply would tend to exhibit relatively stable dynamics of adult numbers. In an LL regime, very high adult numbers would still lead to a crash in adult numbers after about 20 days, largely due to the density dependence of female fecundity. However, if adult numbers were very low, many eggs would be laid but would not cause as large a pulse in recruitment into the adult stage as would be seen in the HL food regime, owing to density-dependent larval mortality in the LL food regime. Similarly, moderate numbers of adults laying eggs in an LL food regime would not elicit levels of larval competition as severe as those in an LH food regime because of density-dependent fecundity in the LL regime. Overall, then, an LL food regime would be predicted to result in adult dynamics that were relatively stable in terms of decreased amplitude of oscillations in adult numbers. Such an effect was, in fact, seen in an experiment in which populations were first maintained on an LH food regime (i.e., larval food supply was held at 50 g per day for about a year and a half). These populations showed the approximate 40-day cycles typical of the LH food regime, with maxima of 2000 to 4000 adults. After a year and a half, the populations were switched to an LL food regime by also restricting the supply of liver to the adults to 1 g per day. This switch in food regime led to a noticeable alteration in the dynamics of adult numbers. The mean number of adults increased three- to fourfold, and although numbers continued to fluctuate, the fluctuations were no longer regular and their amplitude was much reduced (fig. 4.6). Moreover, eggs laid at all adult densities, even fairly high ones, did result in appreciable recruitment into the adult stage under the LL food regime (see fig. 4.6).

As we mentioned earlier, Nicholson also subjected populations maintained on LH or HL food regimes to various regular demographic and environmental perturbations, (Nicholson, 1957). His major interest was in probing the mechanisms whereby populations could compensate for the effects of such perturbations on the numbers of different life stages. One series of experiments involved culling different fixed proportions (50 to 99 percent, depending on treatment) of the adult population every two days in populations subjected to LH or HL food regimes. The major observation from populations subjected to HL food regimes was that the removal of adults essentially reduced competition among females for the limiting protein supply, causing the birth rate per individual to increase. Because there was excess food for larvae in these food regimes, pupal production and adult eclosion per day also increased. To a large degree, therefore, the removal of adults from the cultures resulted in increased recruitment into the adult life stage, partly compensating for the perturbation. For example, the

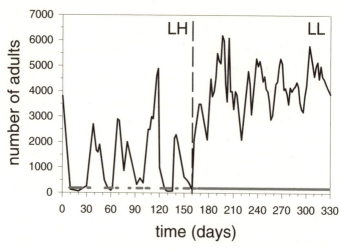

F$_{IG}$. 4.6. Time-series data for 330 days from an experiment in which a population previously kept on an LH food regime (50 g food per day for larvae and excess food for adults) was switched to an LL food regime (adults given only 1 g liver per day). The transition between food regimes is indicated by the vertical dashed line. Horizontal gray bars represent egg-laying periods for which the eggs gave rise to an appreciable number of eclosing adults. Data for the whole experiment have not been shown: The population was maintained for a year after switching food regimes, during which the dynamics were qualitatively similar to what is depicted here for the first 170 days after the switch (modified after Nicholson, 1957).

mean number of adults in cultures kept on an HL food regime without adult removal was 2520, whereas in cultures in which 50 percent of adults were removed every two days, the mean number was 2335; even those cultures in which 90 percent of the adults were removed every two days had a mean adult population as high as 878 (Nicholson, 1954a). In cultures maintained on the LH food regime, an increased removal of adults reduced the numbers of eggs laid per day, thereby alleviating larval competition for food. The reduction in larval competition decreased the mean larval mortality from approximately 98 percent, for cultures in which no adults were removed, to approximately 77 percent, for cultures in which 95 percent of the adults were removed every two days, resulting in increased recruitment into the adult population.

Perhaps more pertinent to our focus on population dynamics and stability are results from experiments involving regular fluctuations in the amount of food provided to adults in populations subjected to HL food regimes. In these experiments, Nicholson (1957) set up 10 populations in which there was an unlimited supply of larval food and water and sugar for the adults. Two of these populations served as controls, with a constant supply of liver to the adults fixed at either 0.1 g or 0.4 g per day. In the other eight populations, the adult supply of liver per day was varied systematically from 0.05 g through 0.5 g and back again. Populations subjected to this fluctuating supply of liver differed from one another in the periodicity (10, 20, . . . 70, 80 days) of the imposed cycles in food supply. These populations were maintained for about two years, and demographic records were maintained as described earlier for the LH and HL regimes.

In all but one of the populations subjected to a fluctuating adult food supply, the dynamics of adult numbers underwent changes in periodicity as time progressed. In the population with adult food supply cycling every 20 days, adult numbers during the first 400 to 500 days fluctuated with the expected periodicity of about 40 days, such that one cycle of adult numbers encompassed two cycles of food supply. Thereafter, however, there was a sudden and dramatic change in the periodicity of the fluctuations in adult numbers, which began to cycle with a periodicity of about 20 days. At the other extreme, adult numbers in the population for which food supply cycled every 80 days fluctuated with the expected periodicity of about 40 days for the first 300 days or so, showing two peaks of adult numbers per food supply cycle. After about 300 days, however, the cycles in adult numbers appeared to break up such that there was only one major peak in adult numbers, with several much smaller peaks, during a single cycle of food supply. Similar effects were seen in other populations with a fluctuating adult food supply; typically the periodicity of the cycles in adult numbers was altered such that it became the same as, or a multiple of the food supply cycle. Nicholson interpreted these results as supporting his view that when a population underwent oscillations in density due to intrinsic factors, environmental cycles would be "impressed upon the population." Why exactly he expected this to happen is not entirely clear, but it is likely that he had in mind some kind of analogy with oscillator entrainment mechanisms.

Interestingly, the dynamics of adult numbers in the control populations in this experiment also underwent a marked change around 400 days after the cultures were initiated (fig. 4.7). During the first 400 days, the control populations exhibited typically large amplitude cycles in adult numbers with a periodicity of about 40 days, as expected in an HL food regime. Thereafter, the fluctuations became very irregular, and the minima in population sizes in particular became much higher than expected in an HL regime. The extremely episodic pattern of egg production typical of the HL food regime, with eggs being laid only during a trough in adult numbers, also broke down after about 400 days. Although the number of eggs laid tended to fluctuate in an inverse relationship to adult numbers, a few hundred or more eggs were being produced at any given time.

Nicholson (1957) interpreted this trend as an indication that evolutionary changes had taken place in these populations over the course of his experiment, and that these changes affected key life-history traits that affected the dynamics of adult numbers in these populations. To test the hypothesis, he measured female fecundity while varying protein supply. Nicholson compared fecundity in the control and fluctuating food supply populations with the fecundity of wild-caught flies and flies from the ancestral laboratory population. He found that flies from the control population and three of the populations used in the fluctuating food supply experiment had far lower minimum protein requirements for egg laying compared with either the ancestral

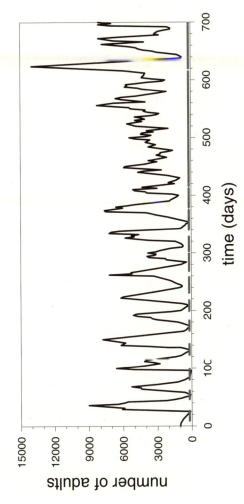

Fig. 4.7. Time-series data from one of the populations that showed an apparent change in dynamics after about 400 days of rearing on a HL food regime. Horizontal gray bars represent times that appreciable numbers (at least a few hundred) of eggs were laid (data from Nicholson, 1957).

population or wild flies. In fact, flies from the experimental populations could actually lay sufficient numbers of eggs to sustain a culture even when given no protein. Nicholson concluded that the extreme competition for protein among adults, at least for part of each food supply cycle, had resulted in natural selection favoring the ability of females to lay eggs despite minimal protein intake.

MODELING THE DYNAMICS OF
LUCILIA CUPRINA POPULATIONS

Given the length of the time series of adult numbers generated by Nicholson's experiments as well as the wealth of demographic information he collected, it is not surprising that these data have been the basis for several modeling efforts, even decades after the experiments were conducted. May (1973) first showed that the time-delayed logistic model, $dN_t/dt = rN[1 - (N_{t-D}/K)]$, the simplest continuous-time population model that can generate cyclic behavior, gave reasonable fits to the observed data with a time delay, D, of about nine days. This exemplifies the inherent problem in post hoc model fitting, because, despite the observed reasonable fit, the time delay of nine days is clearly much smaller than the minimum ontogenetic time delay from egg to egg (about 15 days) in these populations. Similarly, another simplistic model provides reasonably good fits to the observed data (Manly, 1990). In this model, adult mortality rates are compared with adult density at two previous censuses, and recruitment into the adult stage is modeled as a function of adult densities at three previous censuses.

At the other end of the spectrum of model complexity, a stage-, age-, and size-structured model that incorporates mass dynamics also provides good fits to observed dynamics of different life stages (Gutierrez, 1996). These kinds of modeling efforts, however, add little to our understanding of the biological basis of the observed dynamics, especially because there have been no empirical studies on blowflies, subsequent to Nicholson's work, that might enable us to differentiate among these models. Modeling alone, without repeated testing of the models using experimental populations, rarely provides useful insights into factors governing the dynamics of a given population. It is the repeated mutual feedback of theory and experiment that yields great dividends in population ecology research, as we shall see in subsequent chapters on *Tribolium* and *Drosophila*.

An interesting point about blowfly dynamics was made by Brillinger et al. (1980) using a fairly simple model incorporating both age- and density-dependence of adult mortality. Simulations of their model gave rise to chaotic dynamics in which there was a periodic structure for long periods that sometimes broke down into episodes of apparently random dynamics. This result is interesting because of its similarity to results seen in control populations given a fluctuating food supply (see fig. 4.7). In fact, Nicholson often saw the periodic cycles in adult numbers break down over the course of various experiments and typically terminated the cultures because he suspected the cause to be genetic changes in the population. Empirically speaking, it therefore remains an open question as to whether these episodes were in fact a breaking down of the apparently periodic structure in the chaotic dynamics, as was predicted by Brillinger et al. (1980). However, this issue has also been addressed theoretically, as is now described.

A delay-differential equation model, slightly more elaborate than that of May (1973), was used by Gurney et al. (1980) and Nisbet and Gurney (1982) to investigate the mechanisms underlying gross- and fine-level dynamic behaviors seen in the *L. cuprina* populations. They assumed that egg production depends only on current adult density, survivorship from egg to adult depends only on the number of competitors of the same age, and maturation from egg to adulthood takes exactly D time units. They then modeled the adult recruitment rate, R, as $R = R(N_{t-D})$. Further assuming that per-capita adult mortality, δ, is independent of density and age, the rate of change of adult numbers was calculated as $dN/dt = R(N_{t-D}) - \delta N_t$. Finally, recall that in both LH and HL food regimes, recruitment into the adult life stage is essentially zero D time units after a point of high adult density, whereas it is somewhat higher following points of low adult density. Moreover, the recruitment must be zero following a point of zero adult density, and the function has a single maximum. Thus, the dependence of the recruitment function on adult density was modeled as $R(N) = PN \exp(-N/N_0)$, where N_0 is the adult density corresponding to the maximum of the recruitment function. The complete model can, thus, be written as follows:

$$\frac{dN_{t-D}}{dt} = PN_{t-D} \exp\left\{\frac{-N_{t-D}}{N_0}\right\} - \delta N_t \qquad (4.1)$$

Analysis of this model has been described exhaustively by Nisbet and Gurney (1982), so we restrict ourselves here to the major results emanating from the analysis. The model has a single nontrivial equilibrium at $N^* = N_0 \ln(P/\delta)$. The local stability of this equilibrium, and the qualitative aspects of the fluctuations about it, are determined completely by the quantities PD and δD. The observed value of P, the maximal per-capita fecundity, as estimated from Nicholson's data, lies between 7.4 and 11.4 eggs per day. This is consistent with P values required to place the populations in the region of PD-δD parameter space characterized by a stable limit-cycle behavior. The other type of dynamic behavior consistent with observed time-series data is quasicyclic fluctuations, with episodes of relatively regular cycles of period similar to the ontogenetic time delay, D, interspersed with bursts of noise. This kind of behavior can be produced as a result of demographic stochasticity in a system that, from the point of view of a deterministic model, lies in the stable and underdamped region of parameter space (Gurney et al., 1980). Clearly, for this possibility to hold in the case of Nicholson's data, the estimated point equilibrium would need to be stable and underdamped, as opposed to unstable. For the point equilibrium to fall in that region of the relevant parameter space, however, P would need to be an order of magnitude greater than the observed values. It seems reasonably clear, therefore, that the underlying dynamics of the *L. cuprina* populations in the LH and HL food regimes is that of a stable limit-cycle about an unstable point equilibrium. Further support for this conclusion comes from simulations of the model (equation 4.1) under both the quasicyclic and limit cycle hypotheses, with stochastic variation in birth and death rates added to the deterministic-delay differential equation (Renshaw, 1991). The results of these simulations also show clearly that the hypothesis of quasicyclic behavior due to a stable underdamped equilibrium cannot give rise to dynamics that resemble the observed data, whereas simulations under the stable-limit cycle hypothesis capture at least the major qualitative features of the observed dynamics.

Interestingly, deterministic simulations of this model (equation 4.1), with parameter values drawn from Nicholson's data for populations on an LH food regime, are also able to recover the division of adult-stage recruitment into two discrete bursts in each population cycle described earlier (see fig. 4.5). This result is due to the complexity of dynamic behavior exhibited by a population in the limit-cycle region of parameter space especially if it is relatively far from the local stability boundary (Gurney et al., 1980). In this model (equation 4.1), the distribution of adult-stage recruitment within each population cycle depends on how low the adult numbers fall during each trough. When the minimum adult number, N_m, is greater than N_0, recruitment into the adult stage shows one peak per population cycle. When $N_m < N_0$, on the other hand, the single peak begins to show signs of splitting; and finally, when $N_m \ll N_0$, as is the case in the LH food regime, the behavior of recruitment itself becomes cyclic, with two or more peaks of varying degrees

129

of irregularity per population cycle, especially if the cycles in adult numbers are not simple (Gurney et al., 1980).

An extension of the same model (equation 4.1), incorporating the protein-dependence of fecundity, has been used to explain the results from the fluctuating food supply experiment described previously (Stokes et al., 1988). The parameters d, P, N_0, and D were determined from the data on the control population in which adults were given a fixed supply of 0.4 g liver per day. The time series was divided into seven consecutive 100-day periods, and parameters were estimated separately for each such period. The results suggested that, over the course of the experiment, the population moved across the PD-δD parameter space, from the unstable to the stable region, largely as a consequence of reductions in maximal fecundity, P, and the per-capita protein intake at N_0. This latter quantity represents the critical minimum protein requirement for egg laying. Simulations incorporating these time-dependent changes in parameter values were then carried out for conditions mimicking those of the control population (0.4 g protein per day) and those of the population in which adult protein supply fluctuated with a periodicity of 20 days. These simulations were able to capture the changes in dynamics that had been observed in the data after a few hundred generations of maintenance. In the case of the 20-day food supply cycle, when the population was in the unstable region of parameter space yielding stable limit cycles of approximate 38-day periodicity, the effect of the external food supply cycle was merely to drag the period of the intrinsically driven population cycle up to the nearest subharmonic of the food supply cycle (40 days). Once the population had moved into the stable region of parameter space, its intrinsic limit cycle was no longer a constraint, and it began to track the 20-day food supply cycle. The lack of classically stable dynamics in the control population during the last few hundred days of the experiment can be ascribed to noise.

Overall, the results from the delay differential model (equation 4.1) appear to adequately capture various aspects, at both fine and gross levels, of the observed dynamics of *L. cuprina* populations under a variety of environmental conditions. This strengthens our conclusion that the driving force behind these dynamics in adult numbers was the high baseline fecundity and strong time-delayed adult density-dependent recruitment.

CHAPTER FIVE

Tribolium

Flour beetles of the genus *Tribolium* have been used for research in population biology since the early decades of the twentieth century and, along with fruit flies of the genus *Drosophila*, are among the best understood model systems for studying single-species population dynamics. Of the 26 or so species of *Tribolium*, *T. confusum* and *T. castaneum* have been most widely used in population ecology (King and Dawson, 1972). Most of our discussion is therefore limited to these two species. Both species are morphologically and ecologically similar and are easily cultured in the laboratory in coarsely ground flour supplemented with yeast. Chapman (1918) began studies on the biology of *Tribolium* cultures due to its economic importance as a cereal pest, and was soon arguing for its use as a model system to study population ecology. In the 1920s, Chapman was influenced by the theoretical models of population ecology developed independently by Lotka and Volterra, and also studied experimental design with Fisher. These influences coalesced into an approach in which he used careful studies on laboratory cultures of *Tribolium* as a means to estimate empirically the major parameters of the mathematical models, especially those that reflected density-dependent regulatory mechanisms (Chapman, 1928). He also demonstrated major fluctuations in population size when cultures were established only with adults, and argued that cannibalism of immature stages by adults was the major determinant of population size in *Tribolium* (Chapman and Whang, 1934). Chapman's work sparked off a series of long-term studies of the population ecology of *Tribolium* cultures (reviewed in King and Dawson, 1972) that investigated not only the biological aspects of population regulation in single-species cultures—with which we primarily concern ourselves in this chapter—but also some of the earliest studies on the influence of genetic and environmental variation on the outcomes of interspecific competition (Park and Lloyd, 1955; Lerner and Ho, 1961) and studies of dispersal and other behaviors relevant to life-history evolution and population dynamics (Naylor, 1959, 1965; Dawson, 1964; Park et al., 1968).

Indeed, laboratory cultures of *Tribolium* continue to be used extensively for studies in population ecology and evo-lutionary genetics (e.g., Goodnight, 1990b; Wade, 1990; Dennis et al., 1995; Pray, 1997; Benoît et al., 1998). In this chapter, we briefly review the basic biology of *Tribolium* cul-tures and show how this knowledge has been used to develop detailed models of population dynamics that are in close agreement with observed data on population size in lab-oratory cultures of *Tribolium*. Our purpose here is not to cover the biology and laboratory ecology of *Tribolium* species exhaustively: Very detailed and comprehensive reviews can be found in books by Sokoloff (1972, 1974, 1977) and by Costantino and Desharnais (1991). We instead focus on the major determinants of population dynamics in *Tribolium*, in an effort to meaningfully compare and contrast results from different model systems in the final chapter of this book.

LIFE HISTORY OF *TRIBOLIUM* IN THE LABORATORY

Most studies on *Tribolium* cultures have used a protocol that ensures overlapping generations: A census of individu-als in various life stages is taken at regular, often monthly intervals, and the entire population is shifted to fresh medium at each census. In the laboratory, the life cycles of *T. castaneum* and *T. confusum* are similar. Eggs hatch in 4 to 5 days and in 5 to 6 days, respectively, at 34°C and 29°C—the two most commonly used rearing temperatures—with *T. castaneum* eggs hatching somewhat earlier than those of

T. confusum. Typically, cultures are raised on course flour (often with yeast or oil added) in bottles, vials, or other suitable containers kept under constant darkness in incubators with humidities ranging from 20 to 70 percent in various studies. The eggs have a sticky external surface and therefore become coated with the medium. The duration of the larval stage is about 15 to 20 days in these two species, and the number of larval instars varies from 6 to 11 but is usually about 8. The pupal duration is about 5 to 7 days, and females can be fertilized after 3 hours (*T. castaneum*) or 17 to 20 hours (*T. confusum*) after eclosion. Adult females can begin laying eggs at about 100 hours (*T. castaneum*) or 120 hours (*T. confusum*) after eclosion, and adults can live up to 200 days. The life stages can be separated for censusing by passing the medium through a series of sieves with different pore sizes. Variation in the duration of different life stages, as well as in the number of larval instars, is known to be affected by genotype and environmental factors, especially temperature, food, and humidity (King and Dawson, 1972).

Pre-Adult Stages

Compared with *Drosophila* cultures, in which the larvae are the predominant consumers of food resources, the impact of larval density on the dynamics of a typical *Tribolium* culture is not quite so important. In fact, at larval densities typically attained in cultures, the effects of larval density on larval mortality are rather small (fig. 5.1). In typical *Tribolium* cultures, adults coexist with larvae, and thus the primary density-dependent effects on population dynamics are often due to adult rather than larval density. Thus, we must take into account the effects of both larval and adult density on the biology of each life stage. Another unique feature of *Tribolium* population dynamics is the important role of cannibalism in determining population size; both adults and larvae of *Tribolium* eat eggs, smaller larvae, pupae, and callows (very young adults with soft exoskeletons), and at least in

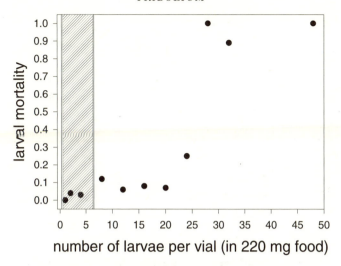

FIG. 5.1. Larval mortality as a function of larval density per unit food medium (number of larvae per vial in 220 mg of food) (data from Howe, 1963). The shaded area represents the range of larval densities (in terms of larvae per unit food medium) typically seen in laboratory cultures of *Tribolium castaneum*. It is clear that for the range of densities seen in typical *Tribolium* cultures, larval density–dependent larval mortality is a negligible factor in determining population dynamics.

larvae, the cannibalism rate increases with age (figs. 5.2, 5.3, and 5.4). Indeed, the major effect of larval density on itself is indirect, through larval cannibalism of eggs. Similarly, the major effect of adult density on larval density is also through the cannibalism of eggs by adults.

Development time from egg to eclosion in *Tribolium* is affected by temperature and humidity. Egg and pupal durations decrease with increasing temperature and are relatively unaffected by humidity, whereas larval duration is affected by both factors. In general, the development of *T. castaneum* is faster than that of *T. confusum*, and this difference is magnified at higher temperatures. Egg-to-adult survivorship of *T. castaneum* is also greater than that of *T. confusum* at higher temperatures, a difference that correlates with the distribution of these two species in the wild (Howe, 1956, 1960).

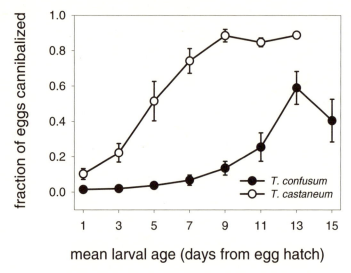

FIG. 5.2. Age-dependent egg cannibalism by larvae of *Tribolium* species (data from Park et al., 1965). Data depicted are the mean (± s.e.) fraction of eggs cannibalized, averaged over four strains of each species. In this experiment, 100 eggs were exposed to predation by 50 larvae of a particular age group, in vials with 8 g food medium.

Development time in *Tribolium* is also affected adversely by increasing adult density, although this is entirely due to a lengthened larval developmental period (Park et al., 1939). A similar effect is seen by manipulating larval density: Increased larval density in an unrenewed medium results in slower larval development and higher larval and pupal mortalities. However, if the medium is renewed at two-day intervals, these density-dependent effects are not seen, suggesting that the direct effects of larval crowding per se are relatively less important than the effects of density on the extent of environmental conditioning. Early work on the effects of larval density on the biology of *Tribolium* has been reviewed extensively by Park (1941).

As the larvae and adults of *Tribolium* feed on and move through the culture medium, they alter its physical and chemical characteristics in many ways. (The food medium

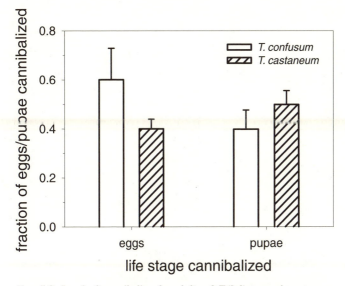

FIG. 5.3. Level of cannibalism by adults of *Tribolium* species on eggs and pupae (data from Park et al., 1965). Data depicted are mean (± s.e.) fraction of eggs per pupae cannibalized, averaged over four strains of each species. Egg cannibalism was recorded in vials containing 8 g of food medium and 100 eggs exposed to predation by 25 males and 25 females for 48 hours. Pupal cannibalism was recorded in vials containing 8 g of food medium and 200 fresh pupae exposed to predation by 10 males and 10 females for 7 days.

in a crowded *Tribolium* culture does not typically disappear, however, as it can in *Drosophila* cultures of high larval density, presumably as a result of high cannibalism of eggs by adults, which tends to keep larval density in check). The nutritive value of the flour is reduced with increasing age of the medium. At the same time, there is a buildup of frass and metabolic wastes as well as gaseous methyl- and ethyl-quinone given off by adults. The flour medium in which adults or larvae have lived for sometime thus acquires a consistency and odor that is referred to in the *Tribolium* literature as "conditioned" (King and Dawson, 1972). Such "conditioning" of the flour medium has no effect on egg hatchability but markedly reduces the extent

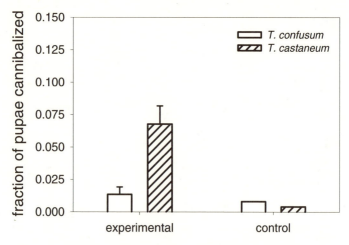

FIG. 5.4. Level of cannibalism by larvae of *Tribolium* species on pupae (data from Park et al., 1965). Data depicted for "experimentals" are mean (± s.e.) fraction of pupae cannibalized, averaged over four strains of each species. Pupal cannibalism was recorded in vials containing 8 g of food medium and 50 pupae exposed to predation by 100 6-day-old larvae for 8 days. Control vials were set up similarly except that no larvae were added; the "control" data are mean levels of pupal mortality in the absence of larval cannibalism.

of adult cannibalism of eggs. It also lengthens the duration of larval development and decreases egg-to-eclosion survivorship (Park, 1941).

There appears to be considerable additive genetic variation for development time in both *T. confusum* and *T. castaneum*, since selection for faster and slower development has been successful in these species (Dawson, 1965; Englet and Bell, 1970). The distribution of development time in *T. confusum* is of the typical bell shape usually seen in insects reared at moderate densities, whereas that of *T. castaneum* is distinctly nonnormal and often bimodal (Lerner and Ho, 1961). There is at present no clear explanation for this difference, but it is interesting in light of a recent observation that adaptation to extremely crowded environments in *Drosophila* laboratory cultures can lead to the evolution of two distinct strategies for dealing with high larval density. Some individuals are fast feeders and early developers, whereas others are slower feeders and developers but exhibit greater tolerance to toxic metabolic waste (Borash et al., 1998).

To sum up, the major effects of density, of both adults and larvae, on the pre-adult stages of *Tribolium* are as follows (some effects of increased density are mediated through the conditioning of food medium):

(1) Pre-adult development time and mortality increase with increasing density.

(2) Mortality of eggs and pupae through cannibalism increases with increasing density of larvae and adults, although cannibalism rates per adult decline with increasing density.

(3) Mortality of eggs and pupae through cannibalism decreases with increased density for any given density of larvae and adults.

(4) At any given density, pre-adult mortality due to cannibalism may be lower in conditioned versus fresh medium.

Of these factors, it is generally felt that the rates of cannibalism of eggs by larvae, and those of cannibalism of eggs and pupae by adults, are the major density-dependent factors affecting the dynamics of *Tribolium* cultures (Dennis et al., 1995) (fig. 5.5).

Adult Stage

Although adults in a typical *Tribolium* culture are outnumbered by larvae and pupae, the adult stage is important in determining the dynamics of these cultures. Adult lifespan is an order of magnitude greater than the duration of the pre-adult stages, and adult density feeds back on pre-adult numbers through conditioning of the medium; density dependence of female fecundity; and, most important, cannibalism of eggs, larvae, and pupae (King and Dawson,

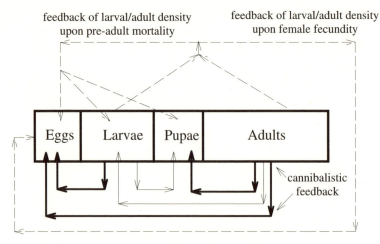

FIG. 5.5. Major density-dependent effects on life-history traits in laboratory cultures of *Tribolium*. Cannibalistic interactions are depicted with solid lines, whereas noncannibalistic effects are depicted with dashed lines. Thick lines indicate effects that are thought to dominate the dynamics of *Tribolium* cultures.

1972; Costantino and Desharnais, 1991). In the previous section(see under *Pre-Adult Stages*), we outlined the major effects of adult density on pre-adult stages. Here we discuss adult cannibalism in slightly greater detail as well as the effect of adult density on female fecundity.

As described earlier, adult *Tribolium* beetles feed on eggs, pupae, and callows. Rates of cannibalism by adults tend to be higher than those of larvae, and adult cannibalism of pupae is more severe than that of eggs (Desharnais and Liu, 1987) (see fig. 5.3). Females of both *T. confusum* and *T. castaneum* exhibit cannibalism rates several times greater than those of males (King and Dawson, 1972). Unlike the rates for larvae, cannibalism rates of adults do not seem to increase significantly with age. The basic nature of cannibalistic interactions in *Tribolium* has been viewed as the outcome of random collisions between eggs/pupae and larvae/adults, with some probability of an egg being eaten on a collision (Crombie, 1946). This theory has led to the formulation of a linear function in which mortality due to cannibalism equals the number of predator (larvae or adult) individuals that is simply added on to an intrinsic mortality rate (Hastings and Costantino, 1987). Alternatively, one can define c_{ij} as the probability that a predator (larva or adult) j encounters a prey (egg or pupa) i in a given period of time, given a total predator population of N during that time interval. Assuming that predator-prey contacts are random and that a contact means the prey gets eaten, the probability of a prey item not getting eaten in the given time interval is $(1 - c_{ij})^N$, which can be approximated as $\exp(-c_{ij}N)$ (Dennis et al., 1995). This approach to modeling cannibalism has been used often in models of *Tribolium* population dynamics, even though it ignores the phenomenon of predator satiation: Survival of pupae despite cannibalism by adults actually increases at higher adult densities (Mertz and Davies, 1968; Park et al., 1968).

In many insect species, the density-dependent control of female fecundity is a major component of population regulation. As with many other insects, the fecundity of *Tribolium* females is dependent on both age and adult density. Fecundity tends to increase rapidly after a few days and peaks relatively early in adult life. Thereafter, it remains around its maximum level for a considerable time before undergoing a more or less linear decrease with age (fig. 5.6). Daily fecundity in *Tribolium* is rather low in comparison with that of many other insects. Even under ideal conditions of a fresh medium, low density, and a temperature of 34°C, fecundity in *Tribolium* does not exceed about 20 eggs per day per female (Park and Frank, 1948). By comparison, the daily fecundity of well-fed *Drosophila* females at very low densities can exceed 100 eggs per day. The decline of fecundity with increasing adult density in *Tribolium* is also not very dramatic. Figure 5.7 shows the results of fitting a hyperbolic equation used to model adult density effects on female fecundity in *Drosophila* ($F(N_t) = a/(1 + bN_t)$) (see Mueller and Huynh,

141

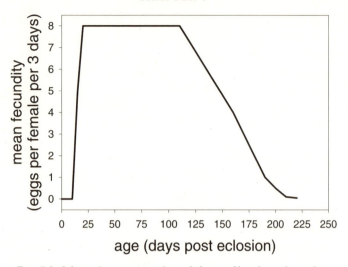

FIG. 5.6. Schematic representation of the profile of age-dependent female fecundity in *Tribolium castaneum* (data from Howe, 1962). Data are from 12 females and 12 males kept in a vial at 25°C and 70% relative humidity. Fecundity increases early in life to a maximum, remains fluctuating at that level until 100 to 120 days, and then undergoes an almost linear decline. This is an idealized representation of observed trends; actual fecundities fluctuate considerably from day to day.

1994) to data on mean fecundity of *Tribolium* females at various adult densities.

The model fits the data well ($R^2 = 0.99$) and yields estimates of $a = 13.61$ and $b = 3.06 \times 10^{-3}$. What is more pertinent here is the low value of parameter b, which determines the sensitivity of female fecundity to increases in adult density. In *Drosophila*, for example, even under a nutritional regime that markedly reduces the sensitivity of fecundity to adult density, the estimate of b is an order of magnitude larger at 2.227×10^{-2} (Mueller et al., 1999). The fecundity of females in *Tribolium* is also reduced in conditioned medium (King and Dawson, 1972). Overall, then, fecundity in *Tribolium* tends to be rather low, even at low densities, and is relatively insensitive to adult density. These two facts together suggest that the density dependence of female

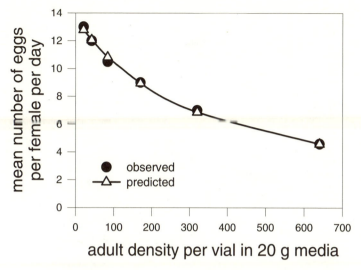

Fig. 5.7. Decline in fecundity of *Tribolium* females with increasing adult density. Observed data are from Rich (1956), pooled across several different treatments with varying durations of exposure to a given density. The solid line connects predicted data points based on a least-squares fitting of a hyperbolic function that models fecundity as a function of adult numbers at time t, N_t : $F(N_t) = a/(1 + bN_t)$.

fecundity may not play a significant role in determining the dynamics of *Tribolium* cultures. Indeed, the two primary determinants of *Tribolium* population dynamics appear to be density-dependent cannibalism rates and rates of adult mortality, which are not strongly density dependent (Dennis et al., 1995). Moreover, adult mortality rates are also relatively age independent, at least over the range of ages for which fecundity is relatively high in *Tribolium* species (fig. 5.8).

A MODEL OF *TRIBOLIUM* POPULATION DYNAMICS

In typical laboratory cultures of *Tribolium* in which food is renewed at regular intervals, larval and pupal numbers tend to show large and fairly regular fluctuations

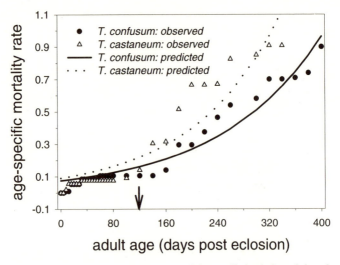

FIG. 5.8. Age-specific mortality rates ("observed" data) for adults of *Tribolium* species (data from Young, 1970). "Predicted" lines correspond to expectations based on fitting the Gompertz equation relating mortality rates to age ($\mu(x) = Ae^{\alpha x}$) to the data from the two species. Parameter estimates are $A = 0.0749$ and $\alpha = 0.00639$ for *T. confusum* ($R^2 = 0.94$), and $A = 0.0901$ and $\alpha = 0.00742$ for *T. castaneum* ($R^2 = 0.87$). The arrow indicates the age beyond which fecundity declines rapidly with increasing age.

whereas adult numbers tend to show relatively stable dynamics (figs. 5.9 and 5.10), consistent with the notion of a steady-state distribution of adult population size (Desharnais and Costantino, 1982; Dennis and Costantino, 1988; Peters et al., 1989). The dynamics of adult numbers are also affected by the initial age composition of the founding population. Populations initiated entirely by adults tend to show greater fluctuations in adult numbers from one census to the next than do populations established with a mix of individuals of the various life stages. Some early studies also established that egg and pupal numbers tend to oscillate in *Tribolium* cultures, largely due to cannibalistic interactions between larvae and eggs (e.g., King and Dawson, 1972). Young (1970) showed that cannibalism of pupae rather than

144

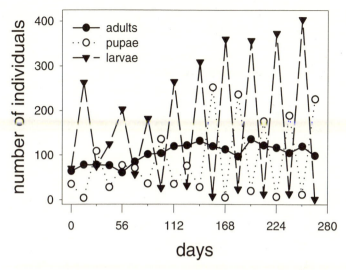

FIG. 5.9. Dynamics of the numbers of larvae (active feeding larvae), pupae (post-feeding larvae, pupae, and callows), and adults over a 20-week period in a culture of *Tribolium castaneum* with resources renewed every 2 weeks. Data are from control population *a* of Desharnais and Liu, 1987, and are fairly representative of the dynamics of *Tribolium* cultures maintained on similar schedules.

eggs was largely responsible for regulating the equilibrium adult number.

A Model of Egg-Larva Dynamics

It is clear from the ecology of *Tribolium* cultures just discussed that the major factors likely to dominate the dynamics of laboratory populations of *Tribolium* are the density-dependent cannibalism of eggs by larvae, and of eggs and pupae by adults. Indeed, the observed oscillations in egg, larval, and pupal numbers can be explained satisfactorily by a consideration of the egg-larva cannibalistic interaction alone. Early on, Chapman (1933) noted the apparent similarity between the phase-lagged oscillations of egg and larval numbers observed in *Tribolium* cultures and those predicted by models of predator-prey interactions.

145

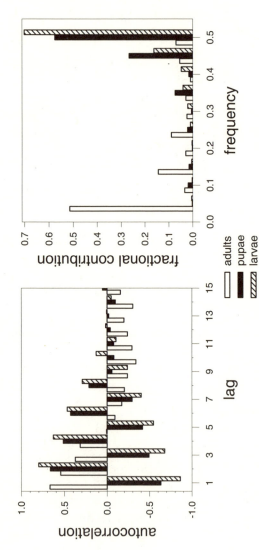

Fig. 5.10. Results of a time-series analysis on data for larval, pupal, and adult numbers depicted in figure 5.9. The first panel shows the autocorrelations, and the second the periodogram of the three time series. All data were detrended before analysis. For larvae and pupae, autocorrelations for the first 7 and 6 lags, respectively, are significant at the 0.05 level, whereas for adults the autocorrelations at lags 1, 2, 10, and 14 are significant.

Hastings and Costantino (1987) focus on the cannibalistic interaction between eggs and larvae, ignoring changes in the number of adults because larval-egg dynamics are taking place on a faster timescale than are adult dynamics. The age (a) distribution in the population at time t, denoted by $n(t, a)$, satisfies

$$\frac{\partial n(t, a)}{\partial t} + \frac{\partial n(t, a)}{\partial a} = -\mu(t, a)n(t, a), \qquad (5.1)$$

where $\mu(t, a)$ is the time- and age-specific death rate, which also encompasses cannibalism rates. Birth rate, $b(t)$, included as a boundary condition $n(t, 0) = b$, is taken as a constant because adult population size is assumed not to change on the timescale of egg-larval dynamics. The death rate, $\mu(t, a)$, is modeled as

$$\mu(t, a) = \mu_e + c_l N_l(t) \qquad \text{for} \qquad 0 < a < A_e \qquad (5.2)$$

and

$$\mu(t, a) = \mu_l \qquad \text{for} \qquad A_e < a < A_e + A_l,$$

where ages 0 through A_e, and A_e through $A_e + A_l$, are assumed to denote eggs and larvae, respectively. Egg mortality through effects other than larval cannibalism is μ_e; this includes egg cannibalism by adults, which is also taken as a constant. The mortality of larvae is μ_l, and the death rate of eggs through larval cannibalism is assumed to be linear and is denoted by $c_l N_l(t)$, ignoring the effect of larval age and egg density on cannibalization rates. Ultimately, a single equation for the number of larvae at time t is obtained:

$$N_l(t) = \int_0^{A_l} b \exp\left[-\mu_e A_e \right.$$

$$\left. -c_l \int_0^{A_e} N_l(t - a - s)ds \right] \exp(-\mu_l a)da \qquad (5.3)$$

Equation 5.3 is the basic model analyzed by Hastings and Costantino (1987), assuming further that the low death rate of *Tribolium* larvae justifies setting $\mu_l = 0$.

Analysis of this model reveals that equation 5.3 has a unique equilibrium as long as $b > 0$. The equilibrium size of the larval population, N_l, is calculated as follows:

$$\hat{N}_l = bA_l \exp(-\mu_e A_e) \exp(-c_l A_e \hat{N}_l) \frac{\left[1 - \exp(\mu_l A_l)\right]}{\mu_l A_l} \quad (5.4)$$

The equilibrium $\hat{N}_l$ can be locally stable or unstable, leading to oscillatory behavior, depending on egg production rate (b), larval cannibalism rate (c_l) and duration of the egg and larval stages (A_e and A_l, respectively). In general, for a given value of c_l and assuming μ_l to be very small, the equilibrium is stable for very short or very long larval periods; this result is relatively independent of the egg production rate (fig. 5.11). For intermediate levels of larval duration (5 to 30 days), the equilibrium is stable only for low egg production (<100 eggs per day). Longer egg durations are destabilizing in this model: As egg duration is increased for any fixed larval duration, the maximal value of b permitting a stable equilibrium decreases dramatically (see fig. 5.11). In addition, increased rates of larval cannibalism are destabilizing, causing a reduction in the parameter space permitting a stable equilibrium.

Of interest in this model (equation 5.3) is that it suggests the presence of biologically meaningful combinations of parameters for which the equilibrium larval number is locally stable. Yet, practically every study on real populations of *Tribolium* species has shown fairly dramatic oscillations in the numbers of eggs, larvae, and pupae (Hastings and Costantino, 1991). By incorporating age-dependent larval cannibalism of eggs into the model described by equation 5.3, Hastings and Costantino (1991) show that the parameter space admitting a stable equilibrium in larval numbers

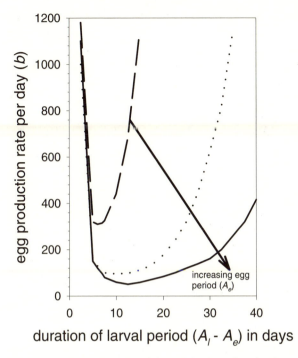

FIG. 5.11. Schematic depiction of the stability boundary for the equilibrium number of larvae in the model of egg-larval dynamics (equation 5.3). The different curves correspond to increasing durations of the egg period. Parameter values above the curves yield unstable equilibria, causing oscillations in larval numbers, whereas those below the curves yield locally stable equilibria (modified after figure 1 in Hastings and Costantino, 1987).

is greatly reduced compared with models assuming a constant cannibalism rate (fig. 5.12).

Indeed, for realistic schedules of age-dependent cannibalism drawn from empirical studies, the value of b required to guarantee a stable equilibrium for given values of the other parameters drops by one to two orders of magnitude. In other words, the incorporation of age-dependent cannibalism of eggs by larvae into equation 5.3 yields the prediction that practically all *Tribolium* populations should show

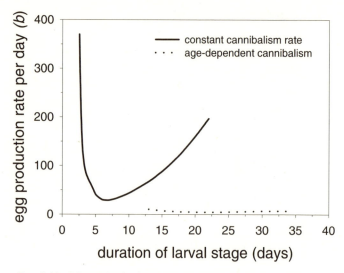

FIG. 5.12. Schematic depiction of the effect of incorporating age-dependent cannibalism of eggs by larvae in the model of egg-larval dynamics (equation 5.3). The stability boundary for the equilibrium number of larvae is depicted for a constant cannibalism rate of 0.024 eggs per larva per day, and for an age-dependent cannibalism rate rising linearly from 0 at age 0 to a maximum of 0.024 at about day 12. Parameter values above the curves yield unstable equilibria, causing oscillations in larval numbers, whereas those below the curves yield locally stable equilibria (modified after figure 3 in Hastings and Costantino, 1991).

sustained oscillations in the numbers of pre-adult stages (Hastings and Costantino, 1991). This model has also been elaborated to include pupal and adult life stages (Hastings and Costantino, 1987), suggesting that increases in adult cannibalism, adult mortality, and pupal duration are stabilizing, whereas increases in fecundity are destabilizing.

Over the years, many different approaches have been used to model *Tribolium* population dynamics. Early on, simple discrete models were used to describe the dynamics of adult numbers in *Tribolium* cultures (e.g., Crombie, 1946; Leslie, 1962). Later, more detailed age-structured models were developed, and various modeling approaches were applied to the question of population growth in *Tribolium*. Detailed reviews of the mathematical models applied to *Tribolium* population dynamics have been provided by Sokoloff (1974) and Costantino and Desharnais (1991). Rather than repeat what has been said before, we restrict our attention to one model of *Tribolium* dynamics—the larva-pupa-adult (LPA) model of Dennis et al. (1995)—that has generated interesting empirical work and holds great promise for future empirical investigation of more complex dynamics in ecology. We also touch on work on *Tribolium* that has been motivated by viewing stable equilibria in population ecology in terms of steady-state distributions of population size. This work, too, has interesting implications for possible future experimental work on demographic stochasticity.

The Larva-Pupa-Adult Model

The (LPA) model of Dennis et al. (1995) consists of three difference equations tracking changes in the numbers of feeding larvae (L); postfeeding larvae, pupae, and callows (P); and mature adults (A), respectively. The model includes density-dependent egg cannibalism by larvae and adults, density-dependent pupal cannibalism by adults, and density-independent fecundity and larval and adult mortality rates. Because the model ignores the egg stage, larvae are considered the stage being recruited. The LPA model is written as follows:

$$L_{t+1} = bA_t \exp\left(-c_{ea}A_t - c_{el}L_t\right) \qquad (5.5a)$$

$$P_{t+1} = L_t(1 - \mu_l) \qquad (5.5b)$$

$$A_{t+1} = P_t \exp\left(-c_{pa}A_t\right) + A_t(1 - \mu_a) \qquad (5.5c)$$

The unit of time in the model is two weeks, which corresponds roughly to the length of the larval duration and half the developmental time from egg to mature adult. Recruitment of larvae at time $t + 1$ is taken to be proportional to

151

the number of adults at time t, shown as A_t. The mean number of larvae recruited per adult in each time interval, in the absence of egg cannibalism, is $b(b > 0)$. The fractions "$\exp(-c_{ea}A_t)$" and "$\exp(-c_{el}L_t)$" are the probabilities of an egg laid between time t and $t+1$ surviving cannibalism by A_t adults and L_t larvae, respectively. The cannibalism of larvae by adults is ignored, and, hence, a fraction $(1 - \mu_l)$ of the larvae at time t become pupae at time $t + 1$. The only cause of pupal mortality is assumed to be cannibalism by adults, and the probability of a pupa surviving to adulthood in the presence of A_t adults is given by $\exp(-c_{pa}A_t)$. Because adult life-span is large, in addition to pupae at time t becoming adults at time $t + 1$, a fraction $(1 - \mu_a)$ of adults at time t survive until time $t + 1$.

This model rests on an understanding of the laboratory ecology of *Tribolium* that has been built up over the years; consequently, only those factors likely to have a large impact on population dynamics enter into the model. Thus, the model ignores the relatively weak density dependence of larval and adult mortality, and of adult fecundity, as well as the limited degree of pupal cannibalism by larvae and larval cannibalism by adults. The age dependence of egg cannibalism by larvae is similarly ignored. Essentially, the LPA model encompasses only density-independent fecundity and larval and adult mortality as well as the density-dependent mortality of eggs and pupae due to cannibalism.

The LPA model (equations 5.5a–c) is not amenable to an analytical description of stability properties unless egg cannibalism by larvae is ignored ($c_{el} = 0$). In this simplified case, there is a trivial equilibrium at extinction ($L, P, A = 0$), which is stable unless $b > \mu_a/(1 - \mu_l)$. If the latter condition holds, there is one nonnegative equilibrium ($L, P, A > 0$), the stability of which depends on fecundity (b), adult mortality (μ_a), and the ratio of the rates of pupal and egg cannibalism by adults (c_{pa}/c_{ea}). Depending on the values of these parameters, the nonnegative equilibrium can be stable or unstable, especially for high values of b and μ_a, giving rise in the latter case to either stable 2-cycles or aperiodic orbits on an invariant loop. Numerical analyses reveal a similar range of dynamic behaviors for the full model that includes egg cannibalism by larvae. A numerically calculated stability portrait of the full model (equations 5.5a–c), incorporating values of c_{ea}, c_{pa}, c_{el}, and μ_l derived by averaging parameter values estimated for four laboratory populations of the *cos* (corn-oil sensitive) strain of *T. castaneum*, shows clearly the dynamic outcomes associated with different regions of $b - \mu_a$ space (fig. 5.13).

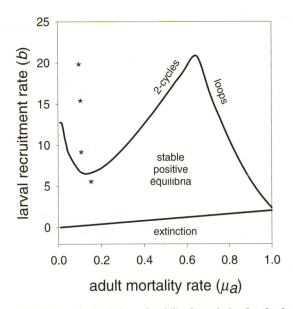

FIG. 5.13. Schematic depiction of stability boundaries for the larva-pupa-adult (LPA) model (equations 5.5a–c) for typical parameter values based on "overall" estimates (see table 5.1) from four laboratory populations of the *cos* strain of *Tribolium castaneum* (modified after figure 3 in Dennis et al., 1995). Asterisks denote the $b - \mu_a$ coordinates of the four populations.

It is clear from this stability portrait that for parameter values in the typical range of *Tribolium* cultures, the most commonly observed dynamic behavior would be stable 2-cycles, although extinction, stable-point equilibria, and loops are also possible (figs. 5.13 and 5.14). Indeed, the LPA model raises the possibility of complex behaviors, especially when adult mortality is relatively high. If fecundity is low (4 to 8 larvae recruited per adult per time interval in the absence of egg cannibalism), there are stable equilibria for a wide range of adult mortality rates. At very high rates of adult mortality, there is a bifurcation from a stable fixed point to an invariant loop, leading to aperiodic cycles. For fecundity values of about 8 to 12, which are often seen in *Tribolium* cultures, a sequence of changes in dynamic behavior takes place as adult mortality rates increase. At extremely low adult mortalities, there are stable equilibria that soon bifurcate to stable 2-cycles. With a further increase in adult mortality, there is a narrow band of μ_a values for which stable equilibria and two-cycles coexist. A continued increase in μ_a again yields a range of values for which only stable equilibria exist, and eventually, at high values of μ_a, a bifurcation to invariant loops occurs.

For values at the higher end of the spectrum of fecundity seen in *Tribolium* cultures ($b = \sim 20$), there are stable 2-cycles until adult mortality is high. Thereafter, there is a zone of multiple attractors: first 2-cycles along with stable fixed points, and later 2-cycles along with invariant loops. For very high values of μ_a, once again the outcome is invariant loops. In practically all situations for which a stable equilibrium exists, this equilibrium is approached via damped oscillations whose amplitudes increase in the case of populations situated near the boundary between stable equilibria and 2-cycles/invariant loops. Thus, for fecundity values of 8 to 22—a range that would include most laboratory populations of *Tribolium*—the stability boundary between fixed

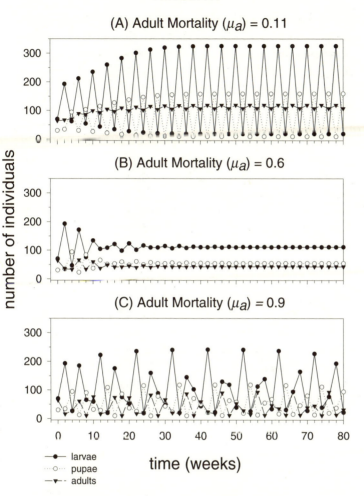

FIG. 5.14. Simulated time series depicting dynamic behaviors corresponding to the three regions in the stability portrait of the larva-pupa-adult (LPA) model (see fig. 5.13). Panels A, B, and C represent parameter values for which the predicted dynamics are stable 2-cycles, stable equilibria, and aperiodic cycles (loops), respectively. All parameter values other than μ_a are the same in all three cases ($b = 11.677$, $\mu_l = 0.5129$, $c_{ea} = 0.011$, $c_{el} = 0.0093$, $c_{pa} = 0.0178$) and are based on "overall" estimates (see table 5.1) from four laboratory populations of the *cos* strain of *Tribolium castaneum* (data in table 5.1, Dennis et al., 1995).

points and 2-cycles would, in practice, tend to be somewhat blurred because of the difficulty in clearly distinguishing between long-lasting transient oscillations of large amplitude and genuine stable 2-cycles.

Another issue of importance to real populations is that of stochasticity in dynamics, whether environmental (due to fluctuations in model parameters as a result of random environmental changes over time) or stochastic (due to intrinsic fluctuations in birth and death rates). The consideration of stochasticity has led to attempts to define and understand population regulation and equilibria in terms of the long-term steady-state distribution of population numbers (e.g., Turchin, 1995a, and references therein). In the case of *Tribolium*, modeling the population of adult numbers using a stochastic variation of the continuous-time exponential model that ignores much of the biology of the pre-adult stages yields $dN/dt = N_t(b\exp[-cN_t] - \mu + \sigma\gamma_t)$; γ_t (Gaussian white noise) is the derivative of a stochastic function (the Wiener increment), which is a continuous-time equivalent of a discrete random variable with no serial autocorrelations, the amplitude of whose fluctuations is measured by σ (Costantino and Desharnais, 1991). From this model, the stationary distribution of adult numbers can be determined in several ways and is approximated by the *gamma* probability distribution (Costantino and Desharnais, 1991). Slightly different formulations also yield the gamma distribution as an approximation of the stationary distribution of adult numbers in *Tribolium* (Costantino and Desharnais, 1981; Dennis and Costantino, 1988; Peters et al. 1989). Similarly, a stochastic version of the LPA model (see equations 5.6a–c, under Empirical Evaluation of the Larva-Pupa-Adult Model) also yields stationary distributions of adult numbers that are well approximated by a *gamma* distribution (fig. 5.15), provided the underlying dynamics are either a stable equilibrium or a stable cycle of relatively small amplitude (Dennis et al., 1995).

Slight changes in the formulation of stochastic models can also result in predictions of stationary distributions of adult numbers that follow, approximately, a normal or lognormal distribution (see fig. 5.15; Desharnais and Costantino, 1982; Dennis and Costantino, 1988). Thus, especially when the "true" model underlying the dynamics of a population is not known, and when several models are in reasonable agreement with the observed data, there is a dubious value of being able to fit a probability distribution to observed frequency distributions of adult numbers in an apparent steady state. Perhaps more interesting is the possibility of using the overall shape of the distribution of adult numbers to determine whether the population is fluctuating about a deterministic stable equilibrium point, or if the fluctuations themselves are deterministic in origin and are merely overlaid by further stochastic noise. The typical prediction in the former case is for a unimodal distribution of adult numbers, skewed to the left; whereas in the latter case, bimodal, multimodal, or irregular distributions may be expected (Dennis and Costantino, 1988). It may be worthwhile to examine whether different types of stochasticity are likely to give rise to different predictions about the nature of the steady-state distribution of adult numbers.

Empirical Evaluation of the Larva-Pupa-Adult Model

There are several complementary ways to evaluate empirically the aptness of a model as a descriptor of the dynamics of real populations. At the simplest and crudest level, one can simply ask how well the model fits observed data. At a slightly more rigorous level, one can examine the predictive power of the model by fitting it to data from populations different from those used to estimate the values of model parameters. An even more rigorous approach is to use the model to predict the dynamic consequences of particular changes in parameter values, and then to test the validity of the predictions using populations exhibiting those specific constellations of parameter values. Such populations

157

can often be obtained through experimental manipulation of the laboratory ecology of the organism, especially in well-studied model systems. In the case of the LPA model, all these approaches have been used, and the overall conclusion is that the model provides a good description of the essential features of the dynamics of laboratory populations of *Tribolium*.

Dennis et al. (1995) evaluated the LPA model using 38-week time series of larval, pupal, and adult numbers from 13 populations of the *cos* strain of *T. castaneum* subjected to different demographic perturbations. These data were from experiments conducted by Desharnais and Costantino (1980), with the complete time series published in Desharnais and Liu (1987). The populations were all initiated with 64 young adults, 16 pupae, 20 large larvae, and 70 small larvae in 20 g of corn oil medium in a half-pint milk bottle. The numbers of adults, larvae, and pupae in each population were recorded every two weeks, at which time all eggs, larvae, pupae, and adults were placed into a fresh culture bottle. Four of the populations served as controls and underwent no perturbation. Of the remaining nine populations, at the tenth week three populations each were subjected to one of three demographic perturbations:

Fig. 5.15. Representative examples of the result of fitting a *gamma* distribution (dotted line) to observed frequency distributions of adult numbers (filled circles) obtained by simulating the larva-pupa-adult (LPA) model with the stochastic components $E_{it}(i = 1, 2, 3)$ assumed to be normally distributed with mean 0 and standard deviation 0.3. All parameter values, other than those indicated on the plots, are listed in table 5.1, "overall." Predicted dynamics for the four cases are stable equilibrium point (A), stable 2-cycles (B, C), and aperiodic cycles (D). In all cases, both the *gamma* and lognormal distributions are consistent with the observed distribution (χ^2 test, $P > 0.35$ in all cases), whereas the normal distribution (the predicted stationary distribution under some stochastic birth-death models) does not fit the data (χ^2 test, $P < 0.005$ in all cases).

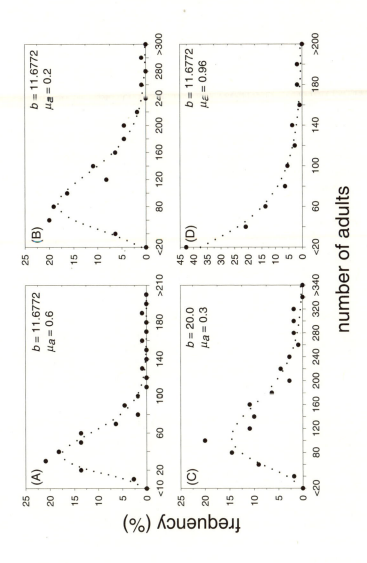

(1) addition of 100 adults, (2) removal of all adults, and (3) removal of all pre-adult stages. To evaluate the aptness of the LPA model, Dennis et al. (1995) fitted a stochastic version of the model to data from the four control populations and obtained both maximum-likelihood and conditional least-squares estimates of the model parameters.

The stochastic version of the LPA model (henceforth, SLPA model) includes noise terms for larval, pupal, and adult numbers that are additive on a logarithmic scale and may be mutually correlated across life stages within a given time interval. The noise terms, however, are assumed to be uncorrelated across time. The SLPA model is, thus, written as follows:

$$L_{t+1} = bA_t \exp(-c_{ea}A_t - c_{el}L_t + E_{1t}) \qquad (5.6a)$$

$$P_{t+1} = L_t(1 - \mu_l) \exp(E_{2t}) \qquad (5.6b)$$

$$A_{t+1} = [P_t \exp(-c_{pa}A_t) + A_t(1 - \mu_a)] \exp(E_{3t}), \quad (5.6c)$$

where $\{E_{1t}, E_{2t}, E_{3t}\} = \textbf{\textit{E}}_t$ is a random vector with a trivariate normal distribution of mean vector $\textbf{0}$ and variance-covariance matrix $\boldsymbol{\Sigma}$. Incorporating error terms in a logarithmically additive manner suggests that environmental stochasticity is assumed to be considerably more significant than is demographic stochasticity (Dennis et al., 1991). To repose greater confidence in the results, parameter values were estimated by maximum-likelihood methods, which are sensitive to departures from multivariate normality of the distribution of $\textbf{\textit{E}}_t$, as well as least-squares methods that are more robust to variation in the distribution of $\textbf{\textit{E}}_t$ and, moreover, should yield estimates similar to the maximum-likelihood estimates if trivariate normality of $\textbf{\textit{E}}_t$ holds good (Dennis et al., 1995).

The parameters for the four control populations, estimated by maximum-likelihood methods (table 5.1), suggest that three of the populations should show 2-cycles whereas one should show a stable point equilibrium (see fig. 5.13).

TABLE 5.1. Maximum-likelihood estimates of parameters of the stochastic larva-pupa-adult model for the four control populations of Desharnais and Costantino (1980)*

Population	b	μ_a	μ_l	c_{ea}	c_{el}	c_{pa}
A	19.85	0.096	0.473	0.016	0.010	0.020
B	15.49	0.100	0.501	0.013	0.010	0.017
C	5.53	0.148	0.508	0.006	0.007	0.018
D	9.13	0.103	0.565	0.009	0.008	0.017
Overall	11.68	0.111	0.513	0.011	0.009	0.018
(95% c.i.)	(6.2–22.2)	(0.07–0.15)	(0.43–0.58)	(0.004–0.018)	(0.008–0.011)	(0.015–0.021)

*The parameter values for the model were fitted to data from all four populations (overall). A 95% confidence interval, calculated from profile likelihoods, for these overall parameters is also given (data from Dennis et al., 1995).

The conditional least-squares and maximum-likelihood estimates were in fairly close agreement, suggesting that the normality assumption for E_t was not grossly violated. The "overall" parameter values, obtained by fitting the SLPA model to data from all four control populations (see table 5.1), also fall into the region of $b - \mu_a$ space where the outcome is stable 2-cycles. The maximum-likelihood estimates of parameter values were then used to fit the LPA model to the data from each of the four control populations. Analysis of the residuals from this fitting for the functions $L_{t+1}(L_t, A_t)$, $P_{t+1}(L_t)$ and $A_{t+1}(P_t, A_t)$ (equations 5.5a–c) also revealed no systematic departure from univariate normality for either of the three state variables L_t, P_t, and A_t (Dennis et al., 1995). The "overall" values were further used to generate one-step forecasts $(E[N_{t+1}|N_t]; N = L, P, A)$ for larval, pupal, and adult numbers at each census period for the four control populations. Comparison of predicted versus observed numbers of individuals showed that these forecasts were reasonably accurate (Dennis et al., 1995). To further test the validity of the LPA model as a descriptor of the underlying population dynamics of *Tribolium* cultures, Dennis et al. (1995) tested the hypothesis that the four populations represented true replications of a single underlying model. This was done by examining, by means of a likelihood ratio test, whether the parameters estimated for the four populations were identical; the test failed to reject the null hypothesis of equality of parameters across the four populations.

To further assess the predictive power of the LPA model, Dennis et al. (1995) used the "overall" parameter values estimated from the four control populations to generate one-step forecasts for larval, pupal, and adult numbers at each census period for the nine populations subjected to demographic perturbations. The question asked here was whether the model could successfully predict the dynamics of populations not used for the parameter estimation. Once again,

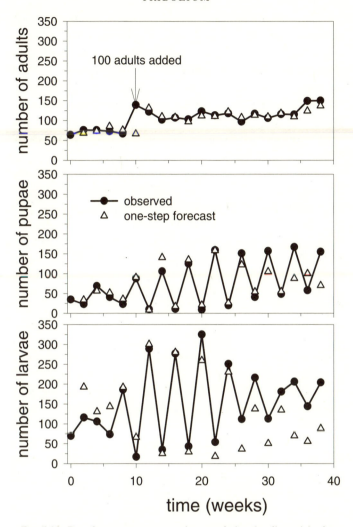

Fig. 5.16. Data from one representative population (replicate A in the treatment wherein 100 adults were added to the culture at week 10 in the experiment of Desharnais and Liu, 1987). Shown are the observed time series for numbers of larvae (active feeding larvae), pupae (post-feeding larvae, pupae, and callows), and adults, along with the one-step forecasts ($E[N_{t+1}|N_t]$; $N = L, P, A$) from the larva-pupa-adult (LPA) model using the "overall" parameter values given in table 5.1.

comparison of predicted versus observed numbers of individuals showed that these forecasts were reasonably accurate (fig. 5.16). In all the three treatments in which populations were demographically perturbed at week 10, the one-step forecasts agreed well with the observed numbers in all replicate populations. In general, the agreement was better for pupal and adult numbers than for larval numbers; this may be a consequence of ignoring age-dependent larval cannibalism of eggs. In the treatment in which all adults were removed, however, there was a major discrepancy between the predicted and observed numbers of larvae at week 12, the census period immediately after the perturbation. The prediction here was of no larvae at week 12, because all adults were removed at week 10. However, as this was not strictly a discrete-time culture, there would still have been eggs in the medium laid by adults between weeks 8 and 10, prior to their removal. Consequently, larvae were observed at the week 12 census, contradicting the prediction. Thereafter, the model again yielded predictions that were consistent with the observed data on larval numbers. The prediction error analysis also supported the LPA model in that only moderate departures from normality were observed for the differences between observed numbers and one-step forecast numbers—and that, too, in only 10 of the 27 time series (Dennis et al., 1995).

Two subsequent studies by R. F. Costantino and coworkers put the LPA model to even more stringent empirical tests (Costantino et al., 1995, 1997; Dennis et al., 1997). In these studies, parameters of the LPA model were estimated empirically for various *Tribolium* strains. The ensuing stability portraits were used to predict the effect of altering adult mortality rate (μ_a) and the rate of pupal cannibalism by adults (c_{pa}) on the dynamics of cultures of these strains. Laboratory populations of these strains were then subjected to experimental manipulation of the values of these two parameters, and the ensuing dynamics recorded and compared to the predictions. In one study (Costantino et al.,

1995), 24 cultures each of two genetic strains of *T. casta-neum* (RR and SS) were established with 100 young adults, 5 pupae, and 250 young larvae in 20 g of media in a half-pint milk bottle. Every two weeks for 36 weeks, the larval, pupal, and adult stages were censused, and all stages (includ-ing eggs) were transferred to a fresh culture bottle. At week 12, four populations of each strain were assigned to each of six treatments that differed in adult mortality rates. In these treatments, adult mortalities of $\mu_a = 0.04, 0.27, 0.50, 0.73$, and 0.96 were experimentally imposed during the census by removing or adding the number of adults required to keep the postcensus number consistent with the assigned mortal-ity rate. Four control populations of each strain, express-ing their intrinsic mortality rate, were also maintained. Data for weeks 12 through 36 from two of the replicate popula-tions in each treatment were used to estimate parameters of the LPA model and to develop stability portraits for use in predicting dynamic behavior at different mortality levels (table 5.2). Data from the remaining two replicates of each treatment were used to evaluate the model predictions.

The various mortality rates were initially chosen so as to place populations in various locations (along the μ_a axis) of parameter space that would yield qualitatively dif-ferent dynamic behavior, based on the stability portrait of the *cos* strain of *T. castaneum* (fig. 5.13) studied by Dennis et al. (1995). However, differences in parameters of the LPA model for the SS and RR strains (table 5.3) gave rise to slightly differing stability portraits, leading to varying predic-tions of dynamic behavior at $\mu_a = 0.75, 0.96$ (see table 5.2).

Overall, the observed dynamics agreed qualitatively with the predictions. At $\mu_a = 0.04$, adult numbers in both strains approached a stable equilibrium quite rapidly (on a timescale similar to that of the controls), whereas larval numbers displayed oscillations for a few weeks longer than the controls before damping became evident. At $\mu_a = 0.27$ and 0.50, for which the prediction was stable 2-cycles, adult

TABLE 5.2. Predicted dynamic behavior of cultures of three genetic strains of *Tribolium casta-neum* at different levels of adult mortality, based on the larva-pupa-adult model

Adult Mortality (μ_a)	*cos* Strain	RR Strain	SS Strain
control	stable 2-cycles	stable equilibrium	stable equilibrium
0.04	stable 2-cycles	stable equilibrium	stable equilibrium
0.27	stable 2-cycles	stable 2-cycles	stable 2-cycles
0.50	stable equilibrium	stable 2-cycles	stable 2-cycles
0.73	stable equilibrium close to boundary for aperiodicities	stable equilibrium close to boundary for 2-cycles	stable equilibrium
0.96	aperiodicities	stable equilibrium close to boundary for aperiodicities	aperiodicities close to boundary for stable equilibrium

TABLE 5.3. Maximum-likelihood estimates of parameters of the larva-pupa-adult model for three genetic strains of *Tribolium castaneum*

Parameter	*cos* Strain	RR Strain	SS Strain
fecundity (b)	11.6772	7.88	7.48
larval mortality (μ_l)	0.5129	0.161	0.267
adult mortality (μ_a)	0.1108	0.0049	0.0036
egg cannibalism by adults (c_{ea})	0.0110	0.011	0.009
egg cannibalism by larvae (c_{el})	0.0093	0.0138	0.0119
pupal cannibalism by adults (c_{pa})	0.0178	0.004	0.004

Data for *cos* strain are from Dennis et al. (1995), and for the RR and SS strains from Costantino et al. (1995).

and larval numbers in both strains displayed regular oscillations that were more pronounced in the case of larvae. Fluctuations in adult numbers at $\mu_a = 0.27$ were regular but of small amplitude and with some indication of damping, whereas at $\mu_a = 0.50$ the adult numbers also displayed sustained oscillations of relatively large amplitude. For $\mu_a = 0.73$, the predictions differed between strains.

The RR strain was in the region of parameter space predicting stable equilibrium but was very close to the boundary of 2-cycles; the populations exhibited sustained oscillations in both adult and larval numbers. The SS strain was predicted to show a stable equilibrium at $\mu_a = 0.73$, and the observed data did suggest a damped oscillatory approach to equilibrium in both larvae and adults. For $\mu_a = 0.96$, the RR strain was predicted to be in the region of stable equilibria but very close to the boundary of the region of invariant loops giving rise to aperiodic cycles. The SS strain, however, was predicted to be in the region of aperiodic cycles but close to the boundary of the region for stable equilibria. Thus, both strains were expected to show aperiodic oscillations, at least for the relatively brief time series observed; populations in the stable equilibrium region close to the boundary of aperiodicities were expected to show aperiodic appearing transients for a fairly long period (Dennis et al., 1995). At a qualitative level, this prediction of aperiodic oscillations was borne out by the data from both strains. Overall, as in the *cos* strain study by Dennis et al. (1995), analyses of time series residuals suggested that the SLPA model provided an adequate description of the dynamics of the RR and SS strains of *T. castaneum* (Costantino et al., 1995).

In another experiment, similar in many ways to the one just described, Costantino et al. (1997) further examined predicted transitions to chaotic dynamics. They used 24 cultures of the RR strain of *T. castaneum*, each initiated with 250 small larvae, 5 pupae, and 100 young adults in 20 g of food medium in a half-pint milk bottle. The adult mortality rate was experimentally set at $\mu_a = 0.96$ for all cultures. All populations were censused and transferred to fresh medium every 2 weeks for a total of 80 weeks. Three populations were assigned to each of eight treatments. In seven of these treatments, rates of recruitment into the adult stage ($P_t \exp(-c_{pa}A_t)$) were experimentally manipulated to yield c_{pa} values of 0.0, 0.05, 0.10, 0.25, 0.35, 0.50, and 1.0. The eighth treatment was a control. Maximum-likelihood estimates of the parameters of the LPA model were used to generate predictions of the dynamic behavior expected in each treatment (table 5.4), and the observed dynamics were compared with the predicted behavior.

In general, the experimental manipulations appeared to have a destabilizing effect. Compared with the control populations, there was a greater degree of fluctuation in the populations subjected to experimentally imposed c_{pa} values. The observed time series, nevertheless, were in reasonably good agreement with the predictions. Lyapunov exponents for chaotic systems are expected to be positive, whereas systems

TABLE 5.4. Predicted dynamic behavior and estimated Lyapunov exponents of laboratory populations of *Tribolium castaneum* subjected experimentally to varying rates of adult recruitment at an experimentally fixed adult mortality level of $\mu_a = 0.96$ (from Costantino et al., 1997; Desharnais et al., 1997)

Pupal cannibalism by adults (c_{pa})	Predicted dynamics	Lyapunov exponent
control (0.0047)	asymptotic approach to equilibrium	−0.0448
0.00	oscillatory approach to equilibrium	−0.2989
0.05	stable 8-cycle	−0.0251
0.10	quasiperiodic behavior (attractor is an invariant loop)	0.0000
0.25	chaotic dynamics	0.0245
0.35	chaotic dynamics	0.1029
0.50	multiple attractors: stable 3-cycle, 8- or higher-period cycles, chaotic attractors	0.0665
1.00	stable 3-cycle	−0.1871

Data for *cos* strain are from Dennis et al. (1995), and for the RR and SS strains from Costantino et al. (1995).

TABLE 5.5. Description of the treatments used by Benoît et al. (1998) in their study of the role of cannibalism in dynamics of laboratory populations of *Tribolium confusum**

Treatment	Expected effect on cannibalism*			Predicted dynamics
	c_{ea}	c_{el}	c_{pa}	
No refuge (control)	1.0	1.0	1.0	Egg-larval (EL) cycles; logistic population growth of adults (A)
Partial refuge for eggs	0.5	0.5	1.0	Stabilization of EL cycles; logistic growth of A
Full refuge for eggs	0.0	0.0	1.0	Rapid stabilization of EL cycles; logistic growth of A
Partial refuge for eggs, larvae, and pupae together	0.5	1.0	0.5	Amplified EL cycles; logistic growth of A to higher equilibrium size
Full refuge for eggs, larvae, and pupae together	0.0	1.0	0.0	Amplified EL cycles; exponential growth of A, at least within duration of the experiment
Partial refuge for eggs, larvae, and pupae separately	0.5	0.5	0.5	Stabilization of EL cycles; logistic growth of A to higher equilibrium size
Full refuge for eggs, larvae, and pupae separately	0.0	0.0	0.0	Rapid stabilization of EL cycles; exponential growth of A, at least within duration of the experiment

*The column on expected effect on cannibalism depicts the fraction to which each cannibalism rate is expected to be reduced by the various refuges: An entry of 1.0 suggests cannibalism at typical levels, 0.5 suggests cannibalism reduced to half its typical level, and 0.0 indicates total protection via refuges, reducing the cannibalism rate to zero. Predicted dynamics are based on simulations of the larva-pupa-adult model with systematic changes in c_{ij} values, keeping other parameters fixed at values reported by Dennis et al. (1995) for the narrain of T culture

with stable equilibria or stable periodic cycles are characterized by negative Lyapunov exponents. Systems with quasi-
periodic invariant loops are expected to show a Lyapunov
exponent of 0. The estimated Lyapunov exponents thus can
be used to numerically categorize the observed dynamics
(see table 5.4).

Benoît et al. (1998) conducted a further empirical study
that evaluated the LPA model but focused on the role of
cannibalism, rather than fecundity and adult mortality, as
a determinant of dynamics in *Tribolium*. They manipulated
cannibalism rates by providing refuges to various life stages
in laboratory populations of *T. confusum*; they then assessed
the role of egg cannibalism by larvae and adults, and of
pupal cannibalism by adults, in determining the long-term
dynamics of these populations. They established 21 populations of *T. confusum*, each initiated with 29 adults and 64
large larvae in 110-mL vials containing 20 g of flour. The
populations were censused every four days and shifted to
fresh culture vials. A total of seven treatments (three populations per treatment), including a control, were imposed on
these populations for 284 days (table 5.5). The dynamics of
egg, larval, and adult numbers expected in each treatment
were predicted by simulations of the LPA model (see table
5.5), although, strangely, these simulations used parameter
values estimated by Dennis et al. (1995) for the *cos* strain of
T. castaneum rather than values for the strain of *T. confusum*
used in the experiments.

In the control populations, no life stage was protected
from cannibalism through a refuge; the levels of cannibalism (c_{ij}; $i = e, p$; $j = a, l$) were therefore expected to be
unchanged from those typically seen in the cultures. Consequently, these populations were expected to show typical
Tribolium dynamics, with sustained egg and larval cycles out
of phase with each other (EL cycles), along with logistic
population growth of adults tending to a stable-point equilibrium (see table 5.5). In the treatments offering refuge to

171

eggs, either half (partial refuge) or all (full refuge) of the eggs in the culture were removed to a separate vial at each census. This protocol was expected to reduce the cannibalism rates on the eggs to 0.5 and 0.0 of the typical (control) values, respectively, resulting in a stabilization of the EL cycles but no change to the pattern of adult growth. Similarly, in the treatments offering refuge to eggs, larvae, and pupae together, half (partial refuge) or all (full refuge) of all three pre-adult stages were removed to one separate vial at each census. These treatments reduced only the cannibalism of eggs by adults and were, therefore, expected to result in amplified EL cycles. Reduction of pupal cannibalism by adults to 0.5 of control values was expected to yield a logistic growth of adult numbers to a stable equilibrium size higher than that of the egg refuge treatments. The complete elimination of pupal cannibalism by adults was expected to free the adult population of density-dependent regulation through the impact of adult density on adult recruitment from the pupal population. Such an effect would lead to a prediction of exponential growth of the adult population, at least for the 284 days of the study. In the last two treatments, either half or all the individuals of each pre-adult life stage were removed to separate vials at each census. These treatments afforded refuge to eggs as well as pupae. They were, therefore, expected to result in stabilization of EL cycles as well as either logistic growth of the adult population to a higher equilibrium size (partial refuge) or exponentially increasing adult numbers (full refuge).

Temporal variability in the observed time series of numbers of small (<2 mm; surrogate for eggs) larvae, large larvae, and pupae was assessed from the standard deviation of log-transformed values of the time series as well as the amplitude of observed cycles. The discrete-time exponential logistic model $A_t = A_{t-1} \exp[r(1 - A_{t-1}/K)]$ was fit to the adult time-series data and used to estimate the equilibrium number K in treatments for which an equilibrium adult population size appeared to be attained. To avoid complications due to the presence of transients, only data after day 125 were used in the analyses. Overall, the effects of the various treatments on the dynamics of small and large larvae, pupae, and adults were in good agreement with predictions based on the LPA model (tables 5.5 and 5.6). The single exception was that pre-adult dynamics in the treatments provided partial or full refuge to all pre-adult stages together in a single container.

The results clearly suggest that adult numbers are controlled primarily by density-dependent feedback acting via pupal cannibalism by adults (see table 5.6). In treatments wherein pupal cannibalism by adults was reduced (partial refuge for pre-adult stages together or separately), adult numbers exhibited logistic growth to an equilibrium value higher than that of controls and treatments with no refuge to pupae. When pupae were given complete refuge, adult numbers grew exponentially for the duration of the experiment, reaching levels of several thousand adults. The numbers of small and large larvae in the control populations exhibited out-of-phase egg-larva cycles characteristic of *Tribolium* cultures, with the amplitude being greater in the case of small larvae. Densities of small and large larvae were, not surprisingly, highest in treatments that provided full refuge to eggs, and were lowest in the controls and in the treatment in which partial refuge was provided to all pre-adult stages together. In the latter treatment, the eggs were exposed to the full strength of cannibalism by larvae and by adults, albeit at a reduced level (see table 5.5). The most stable dynamics, based on reduced amplitude and an inconsistent period of cycles among replicates, were found in treatments providing full refuge to the eggs from cannibalism by larvae and adults (full refuge given to eggs or to all pre-adult stages separately). The amplitude of observed oscillations in numbers of small larvae differed substantially between control populations ($\sim$90% of the mean number) and those given full egg refuge ($\sim$8% of the mean number).

173

TABLE 5.6. Dynamics of pre-adult and adult stages observed in the different treatments by Benoît et al. (1998) in their study of the role of cannibalism in determining the dynamics of laboratory populations of *Tribolium confusum*

Treatment	Observed dynamics
No refuge (control)	Cycles in the numbers of small larvae (SL) and large larvae (LL) out of phase with each other; pupal (P) cycles of smaller amplitude than SL or LL; logistic population growth of adults (A), attaining an equilibrium size of ~90 adults
Partial refuge for eggs	Significant reduction in the amplitude of SL, LL, and P cycles compared with controls; logistic growth of A to an equilibrium of ~90 adults
Full refuge for eggs	Reduction in the amplitude of SL, LL, and P cycles greater but not significantly different from treatment giving partial refuge to eggs; logistic growth of A to an equilibrium of ~150 adults
Partial refuge for eggs, larvae, and pupae together	Cycles in SL, LL numbers of amplitude similar to controls; reduced amplitude of P cycles; logistic growth of A to equilibrium size of ~250 adults
Full refuge for eggs, larvae, and pupae together	Cycles in SL, LL numbers of amplitude similar to controls; reduced amplitude of P cycles; exponential growth of A, reaching ~2500 adults by 284 days
Partial refuge for eggs, larvae, and pupae separately	Significant reduction in the amplitude of SL, LL, and P cycles compared with controls; logistic growth of A to an equilibrium of ~300 adults
Full refuge for eggs, larvae, and pupae separately	Significant reduction in the amplitude of SL, LL, and P cycles compared with controls; exponential growth of A, reaching ~3250 adults by 284 days

Other than the fact that the amplitude of observed cycles was consistently smaller than that of small larvae, the dynamics of large larvae were similar to—and were affected by the various treatments in the same manner as—the dynamics of small larvae.

The effect of different treatments on cycles in pupal numbers was also qualitatively similar to that seen for numbers of small and large larvae. An exception to predicted larval dynamics was seen in treatments providing partial or full refuge to eggs, larvae, and pupae together. In these treatments, eggs would still be subjected to cannibalism by larvae—a factor that induces oscillations in small larval numbers—while being protected from the cannibalism by adults that tends to stabilize egg-larval cycles in *Tribolium*. The prediction for these treatments, consequently, was that the amplitude of cycles in numbers of small and large larvae should rise as compared with that of the controls (see table 5.5). However, the amplitude of small and large larval cycles in these two treatments was of the same order as that of the controls, although the amplitude of pupal cycles was less than that of control populations (see table 5.6). This result is likely due to either (1) egg cannibalism by adults during the four days between each census, at which point individuals were transferred to refuges; or (2) density-dependent reduction of fecundity, because levels of recruitment into the small larval stage remained fairly constant even though adult density changed substantially.

As stated earlier, analytical results from models based on a simple stochastic differential equation, as well as those from simulations of the SLPA model, predict that the stationary distribution of adult numbers in *Tribolium* can be well approximated by the *gamma* distribution. Indeed, in many laboratory populations of *T. castaneum* and *T. confusum*, this has been the case (Dennis and Costantino, 1988; Costantino and Desharnais, 1991). The observed distributions, however, are also frequently consistent with probability distributions other than the *gamma* (Costantino and Desharnais,

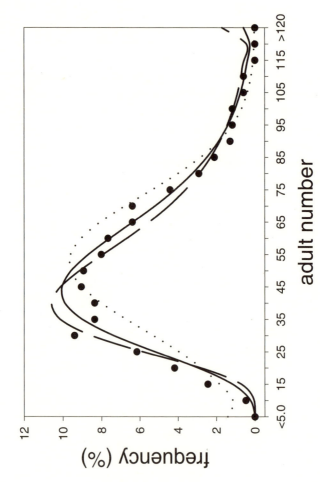

Fig. 5.17. Results from fitting the *gamma* (solid line), lognormal (dashed line), and normal (dotted line) distributions to the frequency distribution of observed steady-state adult numbers (filled circles) in a laboratory population of *Tribolium castaneum* (data from Dennis and Costantino, 1988). The *gamma* and lognormal distributions are consistent with the observed data, whereas the normal distribution is not.

1991); this situation is exemplified in figure 5.17. Moreover, the predicted stationary distribution may vary from species to species, based in part on differences in ecology. In *T. brevicornis*, for example, adults can delay pupation of large larvae, thereby maintaining a relatively constant pool of recruits into adulthood. The predicted stationary distribution for this species is normal, and observations are consistent with that prediction (Desharnais and Costantino, 1982). Thus, predictions about the nature of stationary distributions of adult numbers under different circumstances—and, to an even greater extent, empirical verification of these predictions—are at present somewhat gray areas in which much further work is needed.

Overall, then, it is clear that the dynamics of adult numbers in laboratory populations of *Tribolium* are largely regulated by (1) adult density–dependent cannibalism of eggs, serving to regulate recruitment into the juvenile stage, and (2) adult density–dependent cannibalism of pupae, which in turn regulates recruitment from the juvenile stage into the adult stage. The density dependence of regulation at both these life-stage transitions is fairly strong, leading to relatively stable dynamics of adult numbers for typical laboratory populations. It is also clear that the LPA model of Dennis et al. (1995) provides a good description of *Tribolium* dynamics. Empirical testing of predictions from this model support the view that the dynamics of different life stages in *Tribolium* cultures are largely determined by the interplay of (1) fecundity (all else being equal, higher fecundity is destabilizing); (2) adult mortality; and (3) the rates of cannibalism of eggs by larvae and adults, and of pupae by adults. Indeed, the type of detailed and rigorous empirical work applied to the *Tribolium* model system underscores the importance of laboratory systems to research in population ecology.

CHAPTER SIX

Drosophila

Drosophila has been used as a model organism for research in biology since 1920, and today is one of the best genetically characterized multicellular eukaryotes (Pearl and Parker, 1922). More to the point, *Drosophila* has also been used as a model organism for population ecology since the 1920s, when Raymond Pearl used *Drosophila* to find the "laws" that govern population growth. Pearl took particular interest in trying to understand empirically how intrinsic biological attributes of organisms could lead to density-dependent population regulation. However, much of the enthusiasm for the universality of the logistic model of population growth waned as many of the weaknesses of Pearl's experimental research became known (Sang, 1949). Coincident with and perhaps not because of these developments, the level of research with *Drosophila* in population ecology waned in the 1940s and 1950s. From the 1960s through the 1980s, however, *Drosophila* was again used extensively to study intra- and interspecific competition and population dynamics (Moore, 1952a, 1952b; Miller, 1964a, 1964b; Ayala, 1966, 1969, 1971; Ayala et al., 1973; Gilpin and Ayala, 1973; Barker, 1974; Arthur, 1980, 1986). In the past decade or so, the use of *Drosophila* in population ecology research has again declined, although it is still a useful model system for such work. *Drosophila* has also been used extensively in empirical investigations into the evolution and coevolution of interspecific competitors, a trend that underscores its role in experimental work at the interface of population ecology and evolutionary biology (Moore, 1952b; Futuyma, 1970; Hedrick, 1972; Sulzbach and Emlen, 1979; Joshi and Thompson, 1995, 1996).

In addition to specific findings on the population dynamics of *Drosophila* cultures, much has been learned since 1920 about the important basic biology and laboratory ecology of *Drosophila* (reviewed in Mueller, 1985, 1997). For instance, Chiang and Hodson (1950) established much of the basic laboratory ecology of *Drosophila*. In a monumental study, Bakker (1961) carefully determined the factors affecting competition of *Drosophila* larvae in food-limited environments. These and many other studies have paved the way for a sophisticated and detailed understanding of the effects of food, density, and competitors on important life-history traits in *Drosophila*—a body of information that, in part, makes this species such a useful system for empirical research in population ecology.

In this chapter, we review some of this information and discuss how it can contribute to building a detailed model of population dynamics. Ultimately, this model will be used to make predictions about population stability and life-history evolution. In some cases, these predictions may be tested empirically. The evolution of population growth rates in *Drosophila* (Mueller and Ayala, 1981a) opens the interesting possibility that the stability of populations may also evolve; we review experiments aimed at testing this idea.

LIFE HISTORY OF *DROSOPHILA* IN THE LABORATORY

Drosophila has two active life stages: a nonreproductive larval stage and a reproductive stage as a flying adult. In addition, there is a sedentary pupal stage during which metamorphosis takes place. The larval stage is important to population dynamics for several reasons. First, of course, an individual must survive the larval stage to reproduce. Second, the survival and fertility of adults are affected by the levels of crowding and nutrition they experienced as larvae. We now summarize these effects by considering larvae and adults separately.

179

Larvae

If larvae are crowded into a fixed volume with a constant level of resource, survival and adult size decrease with increasing density (Chiang and Hodson, 1950). These effects, shown in figure 6.1, can be reproduced by keeping the number of larvae constant but decreasing the amount of food available for the developing larvae (Bakker, 1961). One interesting phenomenon, seen in figure 6.1 and in many other studies, is that survival increases rapidly with increasing food level to its maximum value, whereas adult size increases more slowly and reaches its maximum value at food levels far above those needed for maximum survival. This effect is believed to result from the fact that a larva must reach a critical minimum size before it can

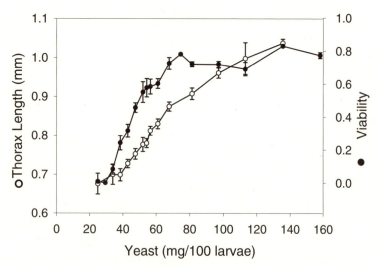

FIG. 6.1. Female thorax length (open circles) and viability (black circles, males and females) of the four *K*-populations described in Mueller et al. (1991a) as a function of food level. Thorax length is highly correlated with adult mass and is used here as a general measure of adult size. The bars are standard errors.

complete metamorphosis successfully, even though larvae typically pupate at sizes much larger than this minimum (Bakker, 1961). As larvae continue to feed beyond this critical point, their additional growth leads to the formation of larger adults, which likely translates into increased female fecundity in the adult stage.

As larvae are crowded, reduced food levels are not the only stress encountered. Burrowing larvae inevitably ingest their own nitrogenous metabolic waste products, largely ammonia, which increase rapidly in crowded larval cultures (Borash et al., 1998). These ingested wastes have toxic effects that reduce survival (Shiotsugu et al., 1997; Borash et al., 1998). Thus, the primary stresses placed on a *Drosophila* larva in a crowded culture are shortage of food and accumulation of nitrogenous waste, both of which tend to intensify with time.

Bakker (1961) demonstrated that *Drosophila* larvae compete for limited food through a scramble type of mechanism. According to this model, larvae may exhibit genetically based differences in rates of food consumption, with the fastest feeders being the superior competitors. This notion was supported by observing that larvae demonstrated as slow feeders could compete successfully if given a head start in feeding (Bakker, 1961). Feeding rates are measured in individual larvae as the number of cephalopharyngeal sclevite retractions per unit time. Competitive ability can be assessed by examining egg-to-adult survival of a particular genotype in the presence of a competitor versus its survival in the absence of the competitor (Nunney, 1983; Mueller, 1988b). Burnet et al. (1977) showed that *Drosophila* larvae whose feeding rates had been increased by artificial selection were also better competitors. The relationship between feeding rates and competitive ability was further clarified by the observation that larvae populations maintained at high densities—whose competitive ability had increased due to density-dependent natural selection (Mueller, 1988a)—had undergone an increase in feeding rates (Joshi and Mueller, 1988); see figure 6.2.

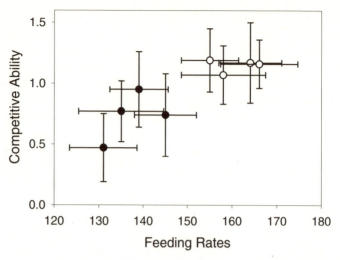

Fig. 6.2. Competitive ability versus larval feeding rates in four K-populations (open circles) and four r-populations (solid circles). The bars are 95% confidence intervals. Feeding rates (Joshi and Mueller, 1988) are quantified by counting the number of times the mouth parts of a larva move back and forth in one minute. Competitive ability (Mueller, 1988a) was measured by assessing the relative increase or decrease in the viability of the wild stock (either the r- or K-populations) in the presence of larvae homozygous for the *white* allele.

In addition to feeding rate, larvae exhibit variation in their foraging behavior in two dimensions. Sokolowski (1980) studied this behavior by quantifying the distance traveled by a larva while feeding on a flat surface. There appears to be a natural polymorphism for foraging path length, controlled by a single locus, *for* (Sokolowski, 1980; de Belle et al., 1989). Sokolowski called those phenotypes that travel little while foraging "sitters," and those that travel greater distances "rovers." Mutants at the *for* locus map to the locus, *dg2*, which encodes a cyclic guanosine monophosphate–dependent protein kinase (Osborne et al., 1997). This protein has been shown to be involved in a variety of nervous system functions (Osborne et al., 1997). Sokolowski et al. (1997) showed that the alleles at the *for* locus respond to density-dependent natural selection, with the rover type becoming common in populations that have evolved at high larval densities and the sitter type predominating in populations kept at low larval densities. However, the precise manner in which foraging path behavior affects fitness components has not been determined.

After *Drosophila* larvae have completed their growth, they search for a place to form their pupal case and complete development, thus bringing the feeding phase of pre-adult life to an end. In the laboratory, pupae generally form on the surface of the food or on the sides of the vials at some distance from the food surface. (This perpendicular distance is typically called pupation height.) The survival of a pupa may be affected by its location (Joshi and Mueller, 1993); see figure 6.3. Under crowded larval conditions, the food becomes a source of mortality for pupae on its surface. Many feeding larvae render the food a soft, semisolid morass, and pupae can get trampled by moving larvae and then drown as they slowly sink into the food. It also appears that mortality is high for pupae located a great distance from the food surface (see fig. 6.3). This leads to a classic form of stabilizing selection in the K populations (see fig. 6.3), whereas selection in the r-populations is predicted to be largely directional since few pupae travel to the highest positions where viability is low (see fig. 6.3). The shape of the selection function is somewhat irregular and certainly not Gaussian in shape.

Adults

If we postpone our consideration of age structure, there are three important remaining determinants of female fecundity: adult size, adult density, and adult nutrition. In general, small females lay fewer eggs than do larger females of the same age (fig. 6.4). Size may account for a roughly threefold difference in fecundity between the very smallest females (thorax length of about 0.6 mm) and the very largest females (thorax length of about 1.1 mm).

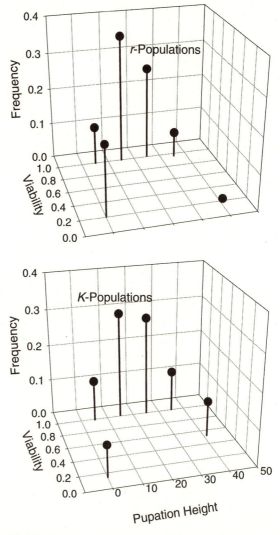

Fig. 6.3. Distribution of pupal heights at high larval density in two populations of *Drosophila melanogaster* (Joshi and Mueller, 1993). The fraction of pupae that survive is also shown. In both populations, viability is very low on the surface and at high pupal heights, but nearly 100% of all larvae survive at intermediate heights. The *K*-populations are less likely to pupate on the surface and tend to pupate at greater distances from the surface of the food than do the *r*-populations.

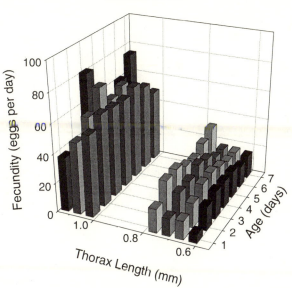

FIG. 6.4. Early female fecundity as a function of female size (thorax length) and age. These data are the averages of eight populations described in Mueller (1987).

Because adult *Drosophila* do not get larger, size differences, which are relics of larval crowding, represent permanent limitations on maximum female fecundity.

The factors other than size that affect female fecundity may be reversed and may vary over time and space. As adult *Drosophila* are crowded, there is a decline in female fecundity, although this is most pronounced in flies that were maintained as adults on low levels of nutrition (fig. 6.5). The combined effects of adult nutrition and adult crowding may cause a fourfold difference in daily female fecundity (see fig. 6.5). The relationship between female fecundity and adult density has an important impact on *Drosophila* population dynamics; we will explore this relationship more fully in the next section (*A Model of Population Dynamics*).

In those cultures in which adults are not segregated from growing larvae, larval density can also have indirect effects on female fecundity through increased levels of nitrogenous

185

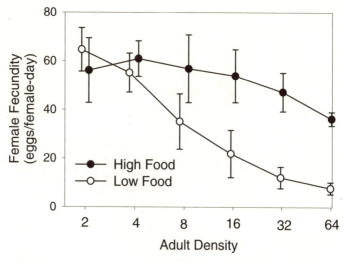

Fig. 6.5. Fecundity of large, young females as a function of adult nutrition and adult density. The bars are 95% confidence intervals. The points are the averages of six populations described in Mueller and Huynh (1994). Fecundity was measured at the six densities shown on the x-axis. Fecundity results at high and low food levels measured at the same density are slightly displaced to ease the visual presentation of these results.

wastes (Aiken and Gibo, 1979; Joshi, Shiotsugu, and Mueller, 1996; Joshi, Oshiro, et al., 1998). The presence of larvae also has direct effects on fecundity. Food medium with larvae at low densities compared with food without any larvae, is preferred as a substrate for oviposition by female *Drosophila* (Del Solar and Palomino, 1966), and high larval densities inhibit fecundity (Chiang and Hodson, 1950).

A MODEL OF POPULATION DYNAMICS

We review the model described by Mueller (1988b), which uses much of the empirical information summarized in figures 6.1 through 6.5. This model uses egg numbers, n_t, as the natural census stage, as discussed in chapter 2. Viability from egg to adult is then assumed to be composed of two parts: A fraction, V, of all eggs is assumed to die due to density-independent causes; the remaining larvae compete for B units of food resources, and survival is density dependent and given by the function $W(Vn_t)$. It is assumed that food is consumed in these environments until it is exhausted, at which point the amount of food consumed by individual larvae differs and follows a normal distribution (fig. 0.0, lower figure) with a mean of $B/(Vn_t)$ and

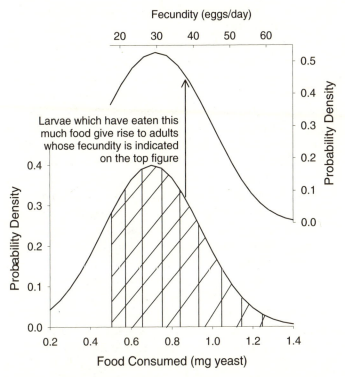

FIG. 6.6. The *Drosophila* population dynamic model. Only larvae that have consumed more than the minimum required food (0.47 mg in the lower figure) survive; the viability of larvae is indicated by the shaded region. The fecundity of the adult females is determined by how much food they eat in excess of the minimum requirement and hence their adult size (see fig. 6.1). The distribution of female fecundity is determined by the distribution of larvae that eat more food than the minimum required.

a standard deviation of $B\sigma/(Vn_t)$. The adult population is drawn from those larvae that have consumed more than the critical minimum (m) amount of food needed to go through pupation successfully. In figure 6.6, these survivors represent the shaded portion of the curve, and the area of this shaded portion is the viability,

$$W(Vn_t) = \int_x^\infty \phi(y)dy, \qquad (6.1)$$

where $\phi(y)$ is the standard normal density function, and $x = (mVn_t B^{-1} - 1)\sigma^{-1}$.

Suppose in the larval population that there are several types (genotypes or sexes) that differ in their relative competitive ability. Let the competitive ability of the *ith* type be α_i. In Mueller (1988b), for instance, there were three types corresponding to the three possible genotypes at a bi-allelic locus. If we know the frequency of each type, we can also define the average competitive ability, referred to as $\bar{\alpha}$. When all the food in the environment has been consumed, the *ith* type will have consumed $\alpha_i B(Vn_t\bar{\alpha})^{-1}$. If the competitive ability of the *ith* type is greater than the population average, these types will consume more food than average, will therefore have a greater chance of surviving, and will, on average, be larger than the rest of the population. It is clear from this formulation that competition is a frequency-dependent process in which benefits to superior competitors accrue only when inferior competitors are present.

Because the surviving larvae have consumed different amounts of food, some are large and some are small, thereby giving rise to adults of varying sizes (see fig. 6.1). From the distribution in equation 6.1, the average size and, hence, fecundity of the surviving females can be computed from the relationships illustrated in figure 6.1. Thus, the mean fecundity of surviving females is a density-dependent function given by the area of the top curve in figure 6.6:

$$F(Vn_t) = W(Vn_t)^{-1} \int_x^\infty f\left[s\left(B(\sigma y+1)V^{-1}n_t^{-1}\right)\right]\phi(y)dy \quad (6.2)$$

$f(\tilde{s})$ is the function that relates adult size, $\tilde{s}$, to female fecundity. For the results in figure 6.6 and elsewhere, $f(\tilde{s})$ is assumed to be a linear function on a log-log scale,

$$f(\tilde{s}) = \exp\!\big(c_0 + c_1 \ln(\tilde{s})\big), \qquad (6.3)$$

where the c_i are empirically determined constants. The size function $s(\tilde{y})$ should increase exponentially with increasing food consumption ($\tilde{y}$) to some maximum value. The function used here is

$$s(\tilde{y}) = a_0 + a_1\{1 - \exp[-a_2(\tilde{y} - m)]\}, \qquad (6.4)$$

where the a_i are also empirically determined constants. However, $a_0 + a_1$ should be the largest female and a_0 should be the smallest female.

The fecundity predicted from $F(\cdot)$ represents the maximum possible given the size of the female. The number of eggs that females actually lay may be further modulated by levels of food provided to the adults as well as the density of adults, as suggested by figure 6.5. The amount of food consumed by adults clearly varies continuously in most populations. Currently, there are no data that can be used to determine the transition that will be taken from one curve in figure 6.5 to the other as food levels vary. For the models that follow, we assume that adult food levels are constant (either very low or at an excess). This assumption is forced on us by our lack of complete information. Ultimately, in our experimental research this assumption can be accommodated since we can easily control the food levels provided to adults in a manner consistent with this assumption.

The effects of adult density on female fecundity can be modeled by a hyperbolic function,

$$G(N_t) = \frac{b_0}{1 + b_1 N_t}, \qquad (6.5)$$

189

where the adult population size, N_t, is given by $W(n_tV)Vn_t$, b_0 is the maximum fecundity reached at low density, and b_1 measures the sensitivity of fecundity to adult crowding. This last variable is ultimately crucial for determining the stability of *Drosophila* populations. Populations that show little sensitivity tend to lay large number of eggs even when populations are crowded, a behavior that tends to destabilize the dynamics of the population. As is shown in figure 6.5, when adults are provided with excess food, their sensitivity to adult crowding is reduced substantially.

We can now combine all these components of the life cycle to produce a recursion in egg numbers:

$$n_{t+1} = \frac{1}{2}G(N_t)F(Vn_t)W(Vn_t)Vn_t \qquad (6.6)$$

The factor of one-half is to account for the fact that only half of the adult population lays eggs. The strength of this model is that the individual components have considerable empirical support. Indeed, many of the parameters of the components parts of equation 6.6 (e.g., those in figs. 6.1, 6.4, and 6.5) can be estimated directly from these experiments (Mueller et al., 1991a). This strength can then be exploited to explore those parts of the life cycle that are most important in determining population stability (see next section, *Stability of Large Laboratory Populations*). However, the liability of this type of model is the large number of parameters it contains, as well as the lack of information about the interaction among these components. As already discussed, for example, the way in which $G(N_t)$ varies with food levels is not known precisely. Moreover, the shape of the hyperbolic functions describing $G(N_t)$ have not been examined for a range of adult sizes. Equation 6.6 implicitly assumes that the effects of adult size and adult food levels act independently on final female fecundity. This same problem was noted with the model developed by Rodriquez in chapter 3.

The natural question that arises is what uses there are for a model such as equation 6.6. It can be used as a means of studying how life histories may evolve. Because the model includes details of the specific life history of *Drosophila*, its predictions can be evaluated directly. Mueller (1988b) concluded that density-dependent selection in *Drosophila* affects competitive ability, which is directly related to larval feeding rates. This prediction has been tested, and the model of viability (equation 6.1) was used to determine the appropriate experimental protocols for measuring competitive ability (Mueller, 1988a). As we explain in more detail in the next section (*Stability of Large Laboratory Populations*), the model also provides a qualitative prediction about the types of environments most likely to result in stable population dynamics of *Drosophila*. Again, these qualitative predictions can be tested easily. However, it is unlikely, even with the parameter estimates obtained from the data previously presented, that numerically accurate predictions of population numbers over time can be obtained from equation 6.6.

Although some progress can be made in determining the analytic conditions for the stability of an equilibrium to equation 6.6 (Mueller, 1988b), these conditions are difficult to interpret. The major determinants of stability are the sensitivity of female fecundity to adult crowding (parameter b_1) and levels of larval food (B). Decreasing b_1 or decreasing B (fig. 6.7) or both tends to move a stable population to cycles and then to chaos. When b_1 is decreased, females tend to lay many eggs, even in crowded environments. Thus, the populations tend to overshoot their equilibrium points. The amount of food provided to larvae can be decreased without decreasing adult resources. Thus, when larval food levels are decreased, even moderate-sized adult populations can overproduce eggs for the meager amount of food available. This again results in the populations overshooting their equilibria and failing to settle to a stable point.

191

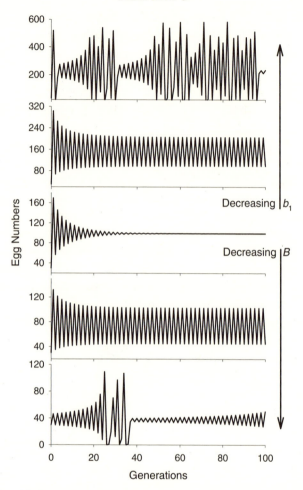

FIG. 6.7. Population dynamic prediction from the *Drosophila* model (equation 6.6). The middle panel shows a stable equilibrium point. As the sensitivity to adult crowding decreases (decreasing b_1), the population moves from a stable point, to a stable cycle to aperiodic behavior (top panels). As the larval resources decrease (decreasing B), the population also becomes progressively less stable (bottom panels). The parameter values for the middle figure were: $B(0.06$ g), $V(0.75)$, σ (0.35 g), $m(0.0003$ g), a_0 (0.5 mm), a_1 (0.623 mm), a_2 (1700 g^{-1}), c_0 (6.041 ln(eggs)), c_1 (2.644 ln(eggs)mm^{-1}), b_0 (1.06), and b_1 (0.3). The other figures differed as follows: top ($b_1 = 0.03$), second from top ($b_1 = 0.1$), second from bottom ($B = 0.03$), bottom ($B = 0.01$).

192

STABILITY OF LARGE LABORATORY POPULATIONS

The theoretical predictions just reviewed can be tested experimentally on laboratory populations of *Drosophila*. Mueller and Huynh (1994) created three different environments that differed in relative amounts of food provided to larvae and adults. In one population, called IIII, the larvae and adults received high amounts of food (fig. 6.8, middle panel). Specifically, adults received excess live yeast and larvae were raised in 40 mL of standard food. A second population, called HL, was maintained as the HH population except that adults received no yeast (fig. 6.8, top panel). The third population, called LH, was maintained as the HH except that the larvae received half as much food (fig. 6.8, bottom panel). The combination of low larval food and high adult food did produce a significant change in population stability, consistent with the model predictions.

To quantify the stability of these populations, the first-order model, the following equation,

$$\ln\left(\frac{N_{t+1}}{N_t}\right) = a_1 + a_2 N_t^{\theta} + a_3 N_t^{2\theta}, \qquad (6.7)$$

was fit to each population, and the best model according to PRESS was used to estimate the stability determining eigenvalue. These populations would not be expected to strictly follow a first-order model. However, since there are so few generations of data, fitting higher-order models is hard to justify. The average of the five eigenvalues is shown in figure 6.8.

This analysis suggests that the LH environment is the least conducive to stable population growth, in accord with the theoretical predictions. Using the least-squares estimates for equation 6.7 for the three populations with eigenvalues less than -1 (LH_1, LH_3, and LH_4), we can observe to what equilibrium the population converges. In each case, iteration of equation 6.7 produced an apparent stable two-point

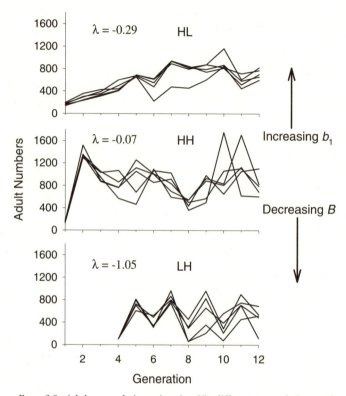

FIG. 6.8. Adult population size in 15 different populations of *Drosophila melanogaster* (from Mueller and Huynh, 1994). The middle panel shows the results for populations kept on high larval and high adult food levels. The populations displayed in the top panel have reduced adult food levels (i.e., increased b_1) relative to the populations in the middle panel. The populations in the lower figure have lower larval food levels (decreased B) than do the populations in the middle figure, and hence have decreased stability. The average of the five eigenvalues (λ) is given for each set of populations. The high level of adult food was simply an excess of live yeast; the low level of larval food was determined by preliminary test that examined several different levels. The theory is not sufficiently precise to predict the low food level accurately.

cycle (LH_1:(154, 761); LH_3:(198, 811); LH_4:(453, 827)). The estimated eigenvalues at these two-point equilibria were also less than one in absolute value in each case (LH_1: 0.49; LH_3: 0.10; LH_4: 0.79). Results from the analysis by the response surface methods are supported by an examination of the autocorrelation function (fig. 6.9). These results also suggest that the LH populations are at an even point cycle.

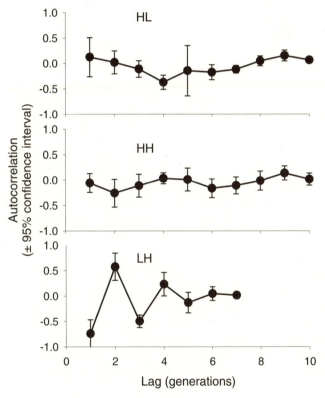

FIG. 6.9. Autocorrelation functions for the populations shown in figure 6.8. Each autocorrelation is the mean of the five replicate populations. The confidence intervals are also based on these five replicates.

STABILITY OF SMALL LABORATORY POPULATIONS

Although the importance of demographic stochasticity in population ecology has been appreciated by theoreticians for at least several decades (e.g., MacArthur and Wilson, 1967; Richter-Dyn and Goel, 1972; Leigh, 1981; Gilpin, 1992), most empirical work on population dynamics has been structured around deterministic models, albeit models of increasing complexity, taking into account specific details of the field or laboratory ecology of the species in question. Empirical results have, by and large, shown reasonable agreement with predictions of the deterministic models. However, the data in such studies have typically been collected from populations large enough to render the effects of demographic stochasticity unimportant to the dynamics of the populations. At the same time, the dynamics of small populations has been receiving considerable attention in ecology in recent years, especially because of the heightened awareness of the need for efficient conservation of biodiversity, much of which is often represented by increasingly smaller populations in an increasingly fragmented landscape (Soulé and Simberloff, 1986; Lande, 1988; Kareiva, 1990; Gilpin and Hanski, 1991, 1997). Much of the theory developed for fragmented populations and metapopulations is also based on simple deterministic models of local subpopulation dynamics. Therefore, it is of considerable interest to assess whether deterministic models of population growth and dynamics can adequately capture at least the essential features of the dynamic behavior of very small populations; or whether we need to explicitly incorporate demographic stochasticity in an appropriate way into our models of population dynamics to make them applicable to smaller populations.

A 1998 study examined the dynamics of eight small populations of *D. melanogaster*, maintained in single 8-dram vials with an average size of 75 adults (s.d. = 57.2) (Sheeba and

Joshi, 1998). Results suggested that the predictions of the *Drosophila* model of Mueller (1988b) regarding the effects of LH and HL food regimes are upheld even for extremely small populations in which demographic stochasticity, acting through fluctuations in sex ratio, birth rate, and death rate, may be expected to be of considerable magnitude. In this study, the linear logistic, exponential logistic, and hyperbolic models were fit to 11 generations of data on adult numbers from 4 populations subjected to an LH food regime, and from 4 to an HL food regime. Both types of populations exhibited fairly large fluctuations in adult numbers, although the coefficient of variation of population size in the LH populations was significantly greater than that seen in the HL populations.

Of the three models fit to the data, only the exponential logistic model gave reasonable fits, with the mean R^2 value for the LH populations (0.65) being significantly greater than that of the HL populations (0.29). Estimates of the intrinsic rate of increase r (table 6.1)—the stability-determining parameter of the exponential logistic model— were consistent with those obtained for larger *Drosophila* populations subjected to LH and HL food regimes (see chapter 2), and also with the qualitative predictions from the *Drosophila* model (Mueller, 1988b). In the HL food regime,

TABLE 6.1. Estimates of the parameter r of the exponential logistic model for eight small populations of *D. melanogaster*, maintained in single 8-dram vials and subjected to HL and LH food regimes (data from Sheeba and Joshi, 1998)

Replicate population	HL food regime	LH food regime
1	1.801	3.076
2	1.156	3.002
3	1.838	3.438
4	1.702	2.294
mean (±95% c.i.)	1.624 (±0.505)	2.953 (±0.761)

all populations exhibited $1 < r < 2$, a condition in which the exponential model predicts an oscillatory approach to a stable equilibrium. In the LH food regime, on the other hand, r in three populations exceeded 3.0, a value for which chaos is predicted, while one population showed $r = 2.294$, for which stable 2-cycles are predicted.

A similar but more detailed study attempted to examine the impact of stochastic variation in sex ratio on the goodness of fit of the exponential model to data on small populations of *D. melanogaster* subjected to either HL or LH food regimes (Joshi, Sheeba, and Rajamani, *unpubl. ms.*). In this study, sets of eight populations each were derived from each of four large (~2000 adults) and outbreeding ancestral laboratory populations. Populations were initiated with eight males and eight females allowed to lay eggs for 24 hours in an 8-dram vial. Four of the populations from each ancestral population were subjected to an LH regime (3 mL food per vial for larvae; yeast supplement for adults), and four to an HL regime (10 mL food for larvae; no yeast for adults). A total of 16 HL and 16 LH populations were set up in this manner.

Each generation, the number of adult males and females present in each population (vial) was counted on day 21 after egg laying. The flies were then placed in a fresh vial with the appropriate amount of food and allowed to lay eggs for exactly 24 hours; the adults were then discarded. The larvae developed and pupated in these vials, and from day 8 through day 18 after egg laying, any eclosing flies in these vials were collected daily into fresh vials containing approximately 5 mL of food. Eclosing flies were added daily into these adult collection vials. Every other day, all adults collected from a specific population until that time were shifted to a fresh vial containing approximately 5 mL of food. On the day 18 after egg laying, the egg vials were discarded and all eclosed adults of each population transferred to fresh vials containing about 5 mL of food (with or without a supplement of live yeast paste added to the wall of the vial, depending on the food regime). Census data on the number of males and females present in each population during egg laying were collected for 11 generations. Once again, both LH and HL populations showed fairly large fluctuations in numbers (fig. 6.10), but the mean (±95% c.i.) coefficient of variation of population size in the LH populations (0.91 ± 0.04) was significantly greater than that seen in the HL populations (0.62 ± 0.04).

Data on the number of adults in each population over 11 generations were subjected to time-series analysis to test the prediction of the *Drosophila* model regarding 2-cycles in the LH regime held true in the face of demographic stochasticity. Linear trends in the individual time series were removed, and autocorrelations estimated between the size of each population at different time lags from one to six generations. Amplitude spectra for each population were also computed, using Fourier analyses on data from generations 3 through 10 for each population. Results of these analyses bear out the prediction of the *Drosophila* model (fig. 6.11). The LH, but not HL, populations exhibit the alternating pattern of negative and positive autocorrelations with an increasing lag that is characteristic of 2-cycles, as well as a distinct peak in the amplitude spectrum corresponding to a frequency of 0.5 (periodicity of two generations).

The sex ratio in the LH populations exhibited large fluctuations from generation to generation with an average coefficient of variation of the fraction of females of 0.24 (95% c.i. = ±0.04). Although there was no systematic departure from a 1:1 sex ratio (fig. 6.12), the fractions of females observed in the LH populations ranged from 0.2 to 1.0. The HL populations, on the other hand, showed a consistently female-biased sex ratio (fig. 6.12), with the mean fraction of females observed at 0.57 (95% c.i. = ±0.2). However, sex ratio in the HL populations was significantly more stable than in the LH populations, varying between extremes of 0.31 and 0.79, with an average coefficient of variation of the fraction of females of 0.16 (95% c.i. = ±0.02).

199

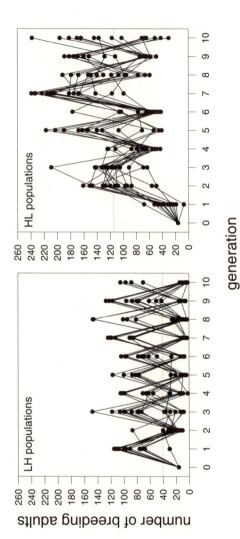

Fig. 6.10. Time-series data on the number of breeding adults in 16 LH and 16 HL small populations of *Drosophila melanogaster* maintained in single 8-dram vials. Dotted lines represent the mean equilibrium number of adults (carrying capacity K in the exponential logistic model), averaged across all populations in each food regime. Mean number of adults ($\pm$ s.d.) was 47.4 $\pm$ 42.1 in LH populations, and 99.6 $\pm$ 59.6 in HL populations.

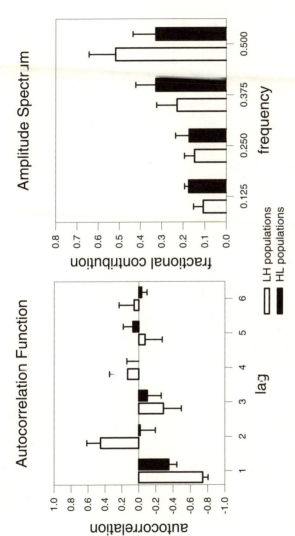

Fig. 6.11. Results of time-series analyses on the data shown in figure 6.10. Plotted values of the autocorrelations and fracional contributions of different frequencies are averaged across the 16 replicate populations within each food regime (LH or HL). Error bars represent 95% confidence intervals about those means.

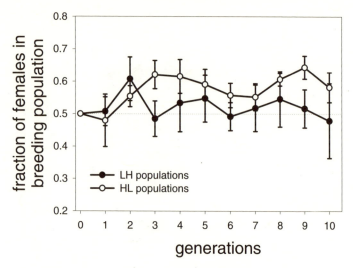

FIG. 6.12. Sex-ratio variation in small HL and LH populations. Data points are the mean fraction of females, averaged across 16 replicates of each food regime. Error bars are 95% confidence intervals about those means.

Fitting of the exponential model to the data from the LH and HL populations gave results similar to those seen by Sheeba and Joshi (1998). To explore the impact of sex-ratio variation on the fluctuations in adult numbers, we fitted the exponential model to the data in two ways, either using adult numbers alone ($N_{t+1} = N_t \exp[r(K - N_t)/K]$) or using twice the number of females instead of the total number of adults ($N_{t+1} = 2Nf_t \exp[r(K - (2Nf_t))/K]$; $Nf_t = $ number of females at generation t). The goodness of fit of the exponential model to the data for both cases, using N_t and $2Nf_t$, was assessed through both the coefficient of determination (R^2), and the mean absolute value of the deviations between observed data and one-step forecasts ($E[N_{t+1}|N_t]$), expressed as a fraction of the mean population size. This measure—henceforth referred to as the coefficient of deviation, D—was calculated as $D = \frac{1}{n} \sum_{i=0}^{n-1} \mathrm{abs}(x_i) / \frac{1}{n} \sum_{i=0}^{n-1} N_i$, where n was the total number of generations, and $x_i = E[N_{i+1}|N_i] - N_{i+1}$.

Mean estimated values of r, K, R^2, and D were obtained by fitting the exponential model to data on total numbers and on twice the number of females. It is clear from these values that the mode of fitting did not have a major impact on the estimates of these various parameters and measures of fit (table 6.2).

Analyses of variance (ANOVAs) confirmed that food regime (LH or HL) had a significant effect on estimates of r, K, and R^2. The pattern of these results was not affected by how the model was fitted to data (i.e., using N_t or $2Nf_t$,) (table 6.3). The pattern of ANOVA results for D was different, however. When model fitting was done using N_t, the effects of block (based on ancestry from a specific population) and the block × food regime interaction were significant. When model fitting was done using $2Nf_t$, none of the ANOVA effects was significant (see table 6.3). The independence of D from food regime is of some interest, because R^2 as a measure of goodness of fit is flawed in this context since it scales with the parameter r. In the LH populations, in which much of the variation in population size is deterministically driven, R^2 is relatively high compared with that of the HL populations in which, presumably, a greater proportion of the variation in population size is due to random fluctuations. The measure D, on the other hand, reflects the average magnitude of deviations from the model's predictions as a fraction of the mean population size.

The dependence of K, R^2, and D on r is examined more finely in figure 6.13. In LH populations, regardless of whether the exponential model was fit using N_t or $2Nf_t$, r and K were negatively correlated ($P < 0.01$). In the HL populations, r and R^2 were positively correlated ($P < 0.001$ when fit used N_t; $P < 0.05$ when fit used $2Nf_t$,). All other correlations were not significant at the 0.05 level. Thus, the scaling of K and R^2 with r appeared to be subject to some

TABLE 6.2. Mean (±95% c.i.) estimates of the parameters r (intrinsic rate of increase) and K (carrying capacity) of the exponential logistic model, and of two measures of goodness of fit (coefficients of determination, R^2, and deviation, D, respectively), for small populations of *Drosophila melanogaster*[*]

| | Estimated from N_t | | Estimated from $2Nf_t$ | |
	LH regime	HL regime	LH regime	HL regime
r	2.937 (±0.233)	1.583 (±0.218)	2.898 (±0.211)	1.682 (±0.112)
K	40.3 (±3.82)	114.9 (±4.18)	40.8 (±2.73)	126.6 (±5.36)
R^2	0.714 (±0.097)	0.283 (±0.113)	0.726 (±0.105)	0.398 (±0.107)
D	0.309 (±0.070)	0.352 (±0.034)	0.315 (±0.063)	0.325 (±0.037)

[*]Populations were maintained in single 8-dram vials and were subjected to HL and LH food regimes. Estimates were made in two ways: (1) using adult numbers alone or (2) using twice the number of females instead of the total number of adults.

TABLE 6.3. Summary of Results of analyses of variance on r, K, R^2, and D estimated for 32 LH and HL populations*

Effect	r		K		R^2		D	
Model Fitted Using N_t	F	P	F	P	F	P	F	P
Block	0.384	0.765	0.005	0.999	1.077	0.378	3.909	0.021
Food Regime	54.664	0.005	738.093	0.0001	19.416	0.022	0.565	0.507
Block × Food Regime	1.475	0.247	0.955	0.430	1.977	0.144	4.987	0.008

Effect	r		K		R^2		D	
Model Fitted Using $2Nf_t$	F	P	F	P	F	P	F	P
Block	1.134	0.355	0.299	0.826	1.193	0.333	1.292	0.300
Food Regime	117.520	0.0017	676.389	0.0001	45.938	0.0066	0.140	0.733
Block × Food Regime	1.020	0.401	1.312	0.293	0.447	0.722	1.055	0.387

*In the body of the table, the F statistic (F) of each test and its significance level (P) are given. Separate analyses were done for estimations based on fitting the exponential model to data using N_t and $2Nf_t$. Food regime (LH or HL) was treated as a fixed factor crossed with random blocks based on ancestry. Degrees of freedom (*df num*, *df denom*) for testing food regime effects are 1, 3; those for the random effects and interactions are 3, 24.

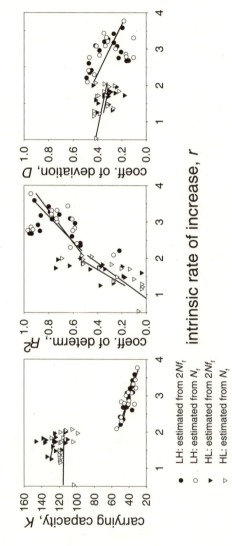

FIG. 6.13. Correlation between the intrinsic rate of increase (*r*), and carrying capacity (*K*), and two measures of goodness of fit (R^2, *D*). Solid lines are least-squares linear regressions through data from each food regime × fitting procedure combination.

limitations based on the range of r values examined; why exactly this is so is not clear.

Overall, the results from this preliminary study of the dynamics of small populations suggest that predictions of the *Drosophila* model (Mueller, 1988b) regarding the dynamic behavior of populations subjected to LH and HL food regimes are borne out even in extremely small populations. It is also clear that the extremely simple exponential logistic model provides reasonable fits to the data (fig. 6.14).

More interestingly, the goodness of fit of this model to these data seems to be unaffected by whether variation in sex ratio was corrected for (see table 6.2 and figs. 6.13 and 6.14), even though sex ratio in the HL populations was consistently female biased (see fig. 6.12). Indeed, the number of adults in generation $t(N_t)$, the absolute deviation of the fraction of females at generation t from 0.5, and the absolute deviation of observed N_{t+1} from the one-step forecasts $(E[N_{t+1}|N_t])$ are mutually uncorrelated (data not shown).

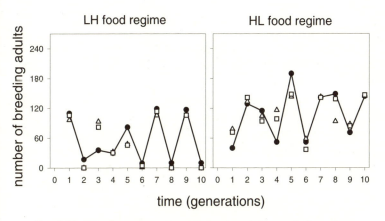

FIG. 6.14. Data from one representative population each from the LH and HL food regimes showing the observed time series of adult numbers (filled circles connected with straight lines), and the one-step forecasts based on two ways of fitting the exponential model to the data: $(E[N_{t+1}|N_t])$ (open triangles) and $(E[N_{t+1}|2Nf_t])$ (open squares).

This result further strengthens the conclusion that sex-ratio fluctuations are not a major contributor to stochasticity in the dynamics of these small populations. Most likely, then, the bulk of the stochasticity is due to variation from generation to generation in birth and death rates, partly due to sampling from a genetically variable population, and partly due to uncontrollable environmental changes. The practical significance of these results is that (1) we can be confident of manipulating even very small *Drosophila* populations through the food regime so as to obtain populations with relatively stable or unstable dynamics; and (2) we can ignore variation in sex ratio while studying gross features of the dynamics of such populations. One important implication of this conclusion is that we can use laboratory systems of *Drosophila* populations to test some of the predictions from theories of metapopulation dynamics and stability discussed in chapter 1. We present results from one such study later in this chapter.

ASSESSMENT OF THE *DROSOPHILA* MODEL

There are two different types of observations that can be used to assess the *Drosophila* model: (1) The quantities that are used to estimate the model parameters can also be used to evaluate the model in a standard goodness-of-fit type of analysis; (2) The model also yields predictions about population level behavior that is independent of the estimation process. If these predictions were accurate, this would constitute strong support for the underlying structure of the model. Under the first category of test, we have examined the model predictions of egg-to-adult viability and average adult size. The model does an admirable job of predicting these quantities (Nunney, 1983; Mueller et al., 1993). In the case of adult size, the model parameters were estimated with one set of experimental data and the predictions made for an independent set of observations (Mueller et al., 1991a).

The second, stronger set of tests involved examination of (1) the variance in female size and (2) stability of the population in a LH environment. We have already reviewed the results of population dynamics in the LH environment for large and small populations. The ability of the *Drosophila* model to correctly predict the qualitative outcome of these experiments is perhaps its most impressive achievement. Mueller et al. (1991a) showed that the variance in female size is correctly predicted by the *Drosophila* model at low food levels but is consistently overestimated at high food levels. In all likelihood, this is due to the assumption that food consumption is normally distributed when in fact there is probably an upper limit to how much food a larva will consume. At high food levels most larvae probably consume close to the upper limit, and thus the variance in size would be expected to decrease.

On the whole the *Drosophila* model yields accurate predictions about many important life history events and population dynamic phenomena. It could be improved by developing a more realistic description of larval food consumption and determining the extent to which there are interactions between larval and adult life histories (e.g., fecundity of different-sized females at different adult densities).

STABILITY IN LABORATORY METAPOPULATIONS

We saw in chapter 1 that various models of metapopulation dynamics differ in their predictions regarding the effect of migration on local (subpopulation) and global (metapopulation) stability. In a metapopulation in which individual subpopulations exhibit relatively large fluctuations in numbers, migration could be destabilizing at the global level by acting as a synchronizing force, tending to bring the various subpopulations in phase with each other. It is also possible, however, that migration, especially if density dependent, could act as a stabilizing force at the local level by damping out fluctuations in individual subpopulations. In this section we describe results from a study in which laboratory metapopulations of *Drosophila* were used to investigate the effect of density-dependent migration on local and global dynamics under differing levels of stability in the local dynamics.

In this study, four metapopulations were initiated from each of four large (~2000 adults) and outbreeding ancestral laboratory populations. Each metapopulation consisted of eight subpopulations maintained in a single 8-dram vial. Each of the four metapopulations derived from a specific ancestor was subjected to a particular combination of stability and migration treatments:

  (1) stabilizing HL type food regime, without migration (SW)
  (2) stabilizing HL type food regime with migration (SM)
  (3) destabilizing LH type food regime without migration (DW)
  (4) destabilizing LH type food regime with migration (DM)

Each of the four treatments was applied to four replicate metapopulations derived from distinct ancestral populations. The experiment, consequently, was of a completely randomized block type with two fixed factors, stability and migration (each with two levels), crossed with blocks based on ancestry. HL and LH food regimes and the maintenance of individual subpopulations were exactly as described earlier in this chapter for the experiment on small population dynamics. Experimentally imposed migration in the SM and DM treatments occurred only among immediate neighbors (fig. 6.15) and was density dependent according to the following arbitrarily determined scheme. Any subpopulation (vial) with fewer than 40 adult flies at the time of census would not contribute emigrants to its neighboring vials. If a subpopulation had 41 to 60 adults, 2 adult females

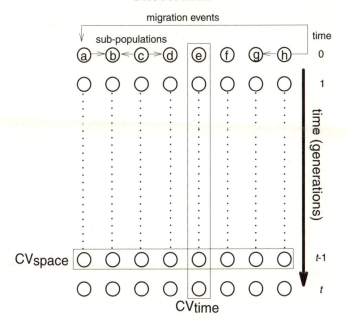

Fig. 6.15. Schematic depiction of the layout of the experimental metapopulations used by Joshi and Sheeba (*unpubl. ms.*). Each row of circles represents a single experimental metapopulation of eight subpopulations (a...h) at a particular generation. Possible immigration and emigration patterns are exemplified by the arrows linking subpopulations in the top row. Each subpopulation, depending on its density, can send out or receive migrants to or from only its immediate neighbors (the array is circularized with respect to migration). The coefficient of variation of adult numbers in a single subpopulation over time (CV_{time}) is a measure of the degree of destabilization of that subpopulation. The coefficient of variation of adult numbers in all eight single subpopulations in a particular generation (CV_{space}) is a measure of the degree of incoherence of the metapopulation at that generation.

would be removed from it after the census. One of these two females would be added to each of the two neighboring vials prior to egg laying for the next generation. A subpopulation with 61 to 80 adults would contribute 2 females to each of its 2 immediate neighbors, and subpopulations with 81 or more adults would contribute 3 females to each neighboring subpopulation. Immigration into a subpopulation was independent of its adult numbers, and the array of subpopulations was circularized for purposes of migration (see fig. 6.15).

The primary interest in this study was to examine (1) the effect, if any, of density-dependent migration on the stability of local subpopulation numbers, (2) the degree of coherence among subpopulations in numbers, and (3) what effects migration may have on the stability of total metapopulation numbers. To this end, the degree of destabilization of a subpopulation was measured as the coefficient of variation of the number of adults in that subpopulation over generations (CV_{time}). Similarly, the degree of destabilization of a metapopulation was measured by the CV_{time} of total adult numbers in the metapopulation over generations. The degree of incoherence of a metapopulation in any generation was measured as the coefficient of variation of adult numbers across all eight subpopulations in that generation (CV_{space}) (see fig. 6.15). Larger values of CV_{space} reflect a situation in which fluctuations in numbers in different subpopulations are more out of phase with each other (greater incoherence).

The fluctuations in the numbers of all types of metapopulations seemed to be reduced in amplitude with time, suggesting that global (metapopulation) dynamics tended to become more stable with time across all four treatment combinations (fig. 6.16, table 6.4). Results from analyses of variance (ANOVAs) on CV_{time} of total adult numbers in metapopulations also supported this conclusion. As expected, D treatments had significantly higher CV_{time}, over the 12 generations of the study, than did S treatments (significant effect of stability: table 6.5, column 1). The stabilization of metapopulation dynamics over time (see fig. 6.16) was reflected in the fact that CV_{time}, over the last six generations of the study, was significantly less than CV_{time} over

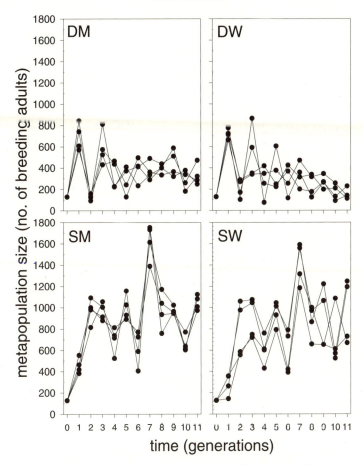

FIG. 6.16. Metapopulation dynamics: time-series data on the total number of breeding adults in each generation, summed over all eight subpopulations in the four types of experimental metapopulations.

the first six generations for all treatments (see table 6.4; significant effect of time: table 6.5, column 2). Moreover, the reduction in CV_{time} from the first six to the last six generations was significantly greater in the case of the D treatments (significant stability × time interaction: table 6.5, column 2).

TABLE 6.4. Mean (±95% c.i.) degree of destabilization (CV_{time}) in the four types of experimental metapopulations*

Treatment	CV_{time} (generations 0–11)	CV_{time} (generations 0–5)	CV_{time} (generations 5–11)
DM	0.439 (±0.16)	0.794 (±0.14)	0.249 (±0.14)
DW	0.615 (±0.16)	0.778 (±0.14)	0.441 (±0.14)
SM	0.355 (±0.16)	0.585 (±0.14)	0.376 (±0.14)
SW	0.389 (±0.16)	0.644 (±0.14)	0.368 (±0.14)

* The mean degree of destabilization is averaged across all four replicate metapopulations within a particular treatment regime. The confidence intervals are based on least-squares estimates of variation among replicate metapopulations, within treatment, in the mixed-model ANOVAs.

TABLE 6.5. Results of analyses of variance (ANOVAs) on the degree of destabilization (CV_{time}) of metapopulations and individual subpopulations within metapopulations*

Effect	CV_{time} for Metapopulations		CV_{time} for Subpopulations	
	Generations 0–11 overall	Generations 0–11 split	Generations 0–11 overall	Generations 0–11 split
Stability	17.56 (0.0248)	11.05 (0.0449)	87.04 (0.0026)	19973.44 (0.0001)
Migration	4.60 (0.1213)	2.84 (0.1908)	20.07 (0.0207)	9.29 (0.0555)
Time	na	1050.25 (0.0001)	na	6.56 (0.0832)
Stability × Migration	1.46 (0.3137)	0.62 (0.4890)	2.18 (0.2367)	0.37 (0.5867)
Stability × Time	na	30.95 (0.0115)	na	41.95 (0.0075)
Migration × Time	na	2.40 (0.2188)	na	1.92 (0.2597)
Stability × Migration × Time	na	3.87 (0.1438)	na	7.04 (0.0768)

* Entries are F values for the tests of various effects, with P values in parentheses. In the ANOVAs done on CV_{time} assessed over the 12 generations of the study (columns 1 and 3), stability (S and D treatments) and migration (M and W treatments) were treated as fixed factors crossed with random blocks. In the ANOVAs done on CV_{time} assessed separately over the first and last 6 generations of the study (columns 2 and 4), stability, migration, and time (generations 0–5 versus generations 6–11) were treated as fixed factors crossed with random blocks. Because each combination of Block × Stability × Migration × Time was replicated only once, random effects and interaction could not be tested for significance and have thus been omitted from the table.

** na, not applicable

Examining the dynamics of individual subpopulations (fig. 6.17) makes it clear that the observed stabilization of metapopulation dynamics was likely due to greater incoherence, as a result of subpopulations drifting out of phase with one another, rather than to any stabilization at the subpopulation level.

In fact, the treatment means for CV_{time} over 12 generations, averaged across subpopulations within a metapopulation and across replicate metapopulations within a treatment, indicate that M treatments may have had slightly more stable dynamics than did W treatments, in which no migration occurred (table 6.6, column 1). This is borne out by the observation of a significant effect of migration in the ANOVA on subpopulation CV_{time}, assessed over the 12 generations of the study (see table 6.5, column 3). Examining the treatment means for subpopulation CV_{time}, assessed separately for the first and last six generations of the study, shows that subpopulations in the D treatments in fact became more destabilized over time, whereas those in the S treatments became slightly less destabilized over time (see table 6.6, columns 2 and 3). Both these effects are significant at the 0.05 level and give rise to a significant stability × time interaction in the ANOVA (see table 6.5, column 4).

It is clear from the preceding results that the S and D treatments had the effect on dynamics that was predicted from the studies of LH and HL small populations discussed earlier in this chapter. It is also evident that migration had no effect on overall metapopulation dynamics, whereas it did have a significant but small stabilizing effect on local subpopulation dynamics.

It also appears that metapopulation dynamics became stabilized over time due to increased incoherence among subpopulations that had drifted out of phase with one another. This last conclusion is strengthened by examining the behavior over time of the degree of incoherence in metapopulations (CV_{space}) in the different treatment combinations.

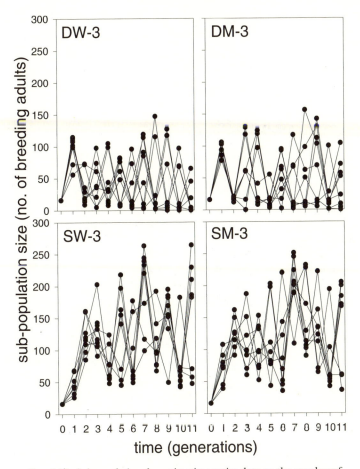

FIG. 6.17. Subpopulation dynamics: time-series data on the number of breeding adults in each generation in each of the individual subpopulations of one representative metapopulation from each of the four experimental regimes. Data from all metapopulations are essentially similar to those depicted.

In both SM and SW treatments, the overall values of CV_{space} were relatively low. After increasing in the first couple of generations, then leveled off and fluctuated within a fairly narrow band thereafter (fig. 6.18). In the DM metapopulations, however, values of CV_{space} increased for the first

TABLE 6.6. Mean (±95% c.i.) degree of destabilization (CV_{time}) in subpopulations of the four types of experimental metapopulations*

Treatment	CV_{time} (generations 0–11)	CV_{time} (generations 0–5)	CV_{time} (generations 6–11)
DM	0.969 (±0.06)	0.948 (±0.16)	1.058 (±0.16)
DW	0.993 (±0.06)	0.876 (±0.16)	1.300 (±0.16)
SM	0.580 (±0.06)	0.612 (±0.16)	0.530 (±0.16)
SW	0.662 (±0.06)	0.729 (±0.16)	0.525 (±0.16)

*The mean degree of destabilization is averaged sequentially across all subpopulations within a metapopulation, and then across all four replicate metapopulations within a particular treatment regime. The confidence intervals are based on least-squares estimates of variation among replicate metapopulations, within treatment, in the mixed-model ANOVA.

six or seven generations before leveling off at values that were about threefold greater than those in the SM and SW treatments. In the DW metapopulations, values of CV_{space} appeared to increase throughout the 12 generations of the study and, toward the last couple of generations, were higher than those seen in the DM metapopulations (see fig. 6.18). Linear regressions fitted separately to data on CV_{space} versus time for each individual metapopulation were all significant at the 0.005 and 0.001 levels for the DM and DW treatments, respectively. In the SM and SW treatments, however, the slopes were much smaller in magnitude (see fig. 6.18), and only two of four metapopulations in each treatment had slopes that were significantly nonzero at the 0.05 level.

An ANOVA on the slopes of these regressions, treating stability and migration as fixed factors crossed with random blocks, yielded significant effects of stability ($P = 0.003$), migration ($P = 0.036$), and the stability × migration interaction ($P = 0.035$). The mean slope in the S treatments was significantly lower than that in the D treatments, and the mean slope in the M treatments was significantly lower than that in the W treatments, in which migration did not occur. The interaction was driven by the fact that the mean slope did not significantly differ between the SW and SM treatments ($P = 0.698$), whereas the mean slope in the DW treatment (0.16) was significantly greater ($P = 0.001$) than that in the DM treatment (0.10).

It is clear from the results of this study that several of the different predictions about the effect of migration on local and global stability in metapopulations seem to hold good. In metapopulations with relatively unstable subpopulation dynamics, increasing incoherence can stabilize dynamics at the global level, and even fairly low levels of migration can be destabilizing at the global level when incoherence is reduced among subpopulations. At the same time, it also appears that migration can play a role in stabilizing the local dynamics by damping out the amplitude of fluctuations in

219

numbers in individual subpopulations. This effect was relatively weak in this study but might be stronger at higher levels of migration. These two effects of migration on the degrees of incoherence and destabilization are contradictory in terms of global metapopulation dynamics, because the former is destabilizing and the latter is stabilizing. Over-

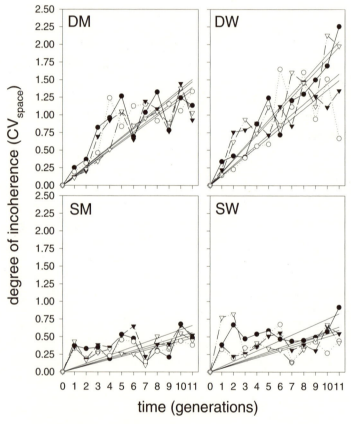

FIG. 6.18. Degree of incoherence (measured as the CV of subpopulation size within a metapopulation at each generation) over 12 generations in each of the experimental metapopulations. Solid lines not connecting any symbols are least-squares regression lines fit to data from each metapopulation.

all, too, migration had no significant effect on metapopulation stability, perhaps because its effects on incoherence and destabilization in part cancel out. This is, to our knowledge, the first empirical study of metapopulation dynamics that attempted to examine the interactions between migration and local and global stability. This is an area in which much theoretical work has been done, and we hope the results from this study will underscore the valuable role that laboratory systems can play in testing predictions from metapopulation theory.

AGE-STRUCTURED POPULATIONS

Many natural populations, including *Drosophila*, consist of adult populations with age structure. There has been extensive research on the evolutionary forces that mold the patterns of age-specific mortality and survival (Rose, 1991; Curtsinger et al., 1995; Mueller and Rose, 1996). However, there has been little work on the dynamics of *Drosophila* (or other species) with age structure, in part because for many species it is difficult to determine the age of individuals. This practical problem has been an insurmountable hurdle for *Drosophila* as well.

We have recently developed methods to overcome this problem and have initiated preliminary experiments to determine the effects of age structure on population stability. The basic problem that we have addressed with these new techniques is the effects of adult age structure on the stability of populations maintained in the LH environments. At this point, we have only suggestive results. However, in this section we describe these techniques and do preliminary analyses of these data, since we consider this such an important problem in population ecology. The techniques used to study age structure could be used with many other insect species and open up new avenues of research in age-structured populations.

221

The basic technique involves painting cohorts of adult *Drosophila* with Testors' enamel paint (diluted with 10% acetone). A small drop is applied to the thorax of the fly with a 0.5-μg syringe (Hamilton, microliter #7000.5). A full syringe provides enough paint to apply to about 20 flies. Experienced painters can paint about 100 flies in one hour. Our preliminary studies of this technique have suggested that male flies painted by experienced people mate as often as unpainted flies in female-choice experiments. In addition, there appears to be no effect of painting on the longevity of adults. The paint does not come off (although if applied to the wings we find the wings rip off quickly), and adults are easily scored. The general protocol for maintaining age-structured populations is outlined in figure 6.19.

The raw data (fig. 6.20) show that the adult population consisted mainly of one-week-old and two-week-old adults. Very few flies made it to the third and fourth weeks of adult life. This was due to fairly high mortality among adults caused by the frequent transfers required by the protocol (see fig. 6.19). This aspect of the environment can be modified easily.

To assess whether these populations are exhibiting any cycling, as we saw in earlier experiments with flies maintained in an LH environment, we performed a spectral analysis on the five populations in figure 6.20 (fig. 6.21). These results suggest that there may be a periodic component at 0.16 cycles per week, which corresponds to a period of six weeks. This periodicity is strongest in the total population size. To interpret this observation, recall that the time units in figure 6.21 are weeks, not generations as in figure 6.9. After the adult population has laid eggs, the first progeny from these eggs are produced two weeks later, although most emerged during the third week. Because the total population size is composed mostly of one-week-old flies, these progeny have a big impact on total adult numbers. Thus, the bust-or-boom cycle seen in figure 6.9 that occurs with a period of two generations would appear in the age-structured populations with a period of six weeks, which we have observed here.

Thus, our preliminary result is that the addition of age structure has not removed the cycling that was present in the

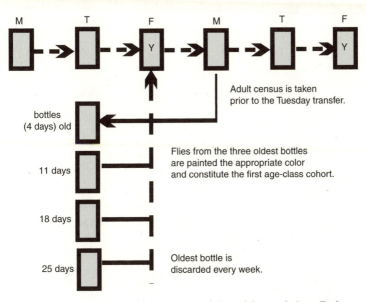

FIG. 6.19. Maintenance of age-structured *Drosophila* populations. Each rectangle at the top represents a half-pint *Drosophila* culture. The letters above each culture stand for a day of the week. Every Monday, the adult population is transferred to a fresh culture to lay eggs. After 24 hours (on Tuesday), the adults are removed and the numbers in each age class counted. The egg-laden culture is then saved for future collections of progeny. On Friday the adult population is again moved to a fresh culture. However, progeny that have emerged from the egg-laying cultures that are 11, 18, and 25 days old (there are never progeny in the 4-day-old culture) are collected, painted one color, and added to the adult population. The dashed lines at the top indicate the transfer of flies from one culture to another. The solid lines represent the movement of an entire culture. This maintenance regime resembles the LH environment in that adults are given excess yeast (indicated by the "Y") before egg laying, and larvae develop in cultures with low levels of food (15 mL).

223

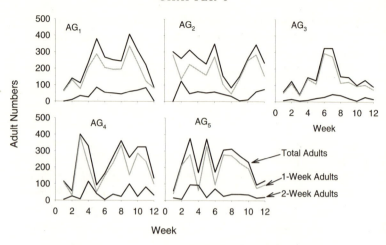

FIG. 6.20. Age-class numbers in five populations of *Drosophila melanogaster* maintained by the protocol outlined in figure 6.19. The first two weeks of the experiment are not included.

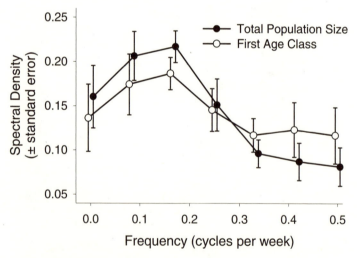

FIG. 6.21. The spectral density function estimates for the five populations shown in figure 6.20. The points are obtained from the average spectral density from each of the five populations. Standard errors are also estimated from these five observations. The series were zeroed and detrended prior to the analysis. A Hamming window was used with five adjacent observations (Chatfield, 1989, p. 116).

LH populations without age structure. The model developed in chapter 2 (equation 2.16) suggests one reason for this result. That model consisted of only two adult age classes, and indeed the populations maintained here (see fig. 6.20) consisted of mainly two adult age classes. If survival from the first adult age class to the second was low, age structure did not produce a stable equilibrium point (see fig. 2.5). In this experiment, the average survival from the first to the second adult age class was quite low: 0.24 ± 0.04 (95% c.i.). Clearly, more work is needed before the role of age structure in determining population stability can be evaluated fully. With this experimental system, we would need to run the experiment for a much longer period so contemporaneous controls (no age structure) could be run to demonstrate that the LH environment does produce cycles. Likewise, the techniques would need to be altered (e.g., by keeping the adults in cages) so that adult survival is increased.

EVOLUTION OF POPULATION DYNAMICS

The evolution of population dynamics has served as one of the earliest unions of ecological and evolutionary theory. MacArthur (1962) introduced the idea that the carrying capacity may be an ecological analog of fitness. These ideas were extended by MacArthur and Wilson's (1967) development of r- and K-selection theory. The synthesis was complete with the development, by several people, of formal population genetic theories of density-dependent natural selection (Anderson, 1971; Charlesworth, 1971; Roughgarden, 1971; Clarke, 1972; Asmussen, 1983).

The most important assumption of these models is that fitness is equivalent to per-capita rates of populations growth. For the simple single-locus versions of these models, we know from population genetic theory that selection maximizes fitness (Kingman, 1961), and hence population growth rates are maximized. In population-growth models, such as the logistic, fitness at high density is closely related to the carrying capacity. We therefore see in one class of models selection resulting in the maximization of the carrying capacity (Roughgarden, 1976). However, there are several other theoretical settings in which selection does not necessarily maximize the equilibrium population size (Prout, 1980).

Ultimately, the consequences of natural selection on population growth rates must be studied empirically. With this goal in mind, we undertook a laboratory study of density-dependent selection in *Drosophila* in 1978. The populations used in this study originated from wild-caught populations from Berkeley, California, in 1975. The history of these populations is outlined in figure 6.22.

In keeping with MacArthur and Wilson's original formulation of density-dependent natural selection, two environments were created to study the evolution of *Drosophila* at extreme densities. The *r*-environment maintained larvae and adults at low density (Mueller and Ayala, 1981a). Adults reproduced during the first week of adult life only. The size of the breeding adult *r*-populations was only 50 for the first 188 generations, after which time the breeding population size was increased to 500. The *K*-populations were kept at high larval and adult densities by culturing the flies with the serial transfer technique (see fig. 3.1). In these populations, the breeding number of adults was close to 1000, and adults were permitted to reproduce until they died. There were three independent replicates of each *r*- and *K*-population.

The first measurements of population growth rates, made after just eight generations, and showed that at low population densities the growth rates were higher in the *r* populations relative to the *K*-populations, but that the reverse was true at high population densities (fig. 6.23). This genetic differentiation may have been due to genetic changes in only the *r*-populations, only the *K*-populations, or in both populations. To sort this out, several new populations were created after 198 generations of *r*-selection (see fig. 6.22).

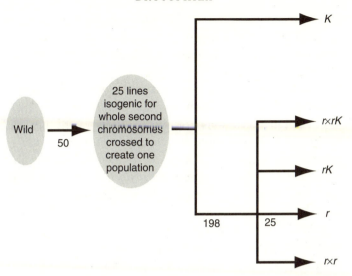

Fig. 6.22. Origin and history of laboratory populations of *Drosophila melanogaster* used to study density-dependent selection. Flies caught in nature were brought back to the laboratory and used to establish 25 new populations that were each homozygous for a different second chromosome from these wild flies (Mueller and Ayala, 1981d). These lines were kept in the laboratory for about 50 generations before they were crossed to create a genetically variable population. This population was used to create three *K*-populations and three replicate *r*-populations. After about 198 generations in the *r*-regime, three new types of populations were created, each replicated threefold. The *r* × *r* populations were created by doing all pairwise crosses of the three *r*-populations. The progeny of these crosses were also used to create the *r* × *rK* populations. The *r* × *r* populations were kept in the *r*-environment, whereas the *r* × *rK* populations were kept in the *K*-environment. The *rK* populations were derived from each of the three r-populations but were maintained in the *K*-environment.

Samples of flies from each of the replicate *r* populations were introduced to the *K* environment, these populations were called *rK*'s. After 25 generations, population growth rates were measured in the three replicate *r*- and *rK*-populations (see fig. 6.23). For this experiment, it was reasonable to assume that the *r*-populations had changed little in those 25 generations, since they had had 198 generations to adapt to the laboratory environment. These results were consistent with the earlier observations; growth rates of the rK-populations were depressed at low density but elevated at the high densities relative to the r-populations (Mueller et al., 1991b).

Because the r-populations had undergone many generations of drift at a much smaller population size than the K-populations, some of these results might be due to the accidental fixation of deleterious mutants in the r-populations, or perhaps the loss of genetic variation in the r-populations. The $r \times r$ populations reintroduced genetic variation into each r-population by performing all possible crosses between the three r-populations. When the $r \times r$ populations were moved to the K-environment and allowed to evolve for 25 generations, changes in population growth rates similar to the r- and rK-populations were observed (see fig. 6.23).

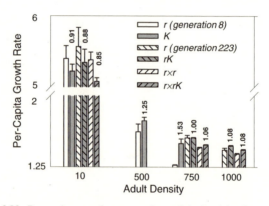

Fig. 6.23. Per-capita growth rates at four adult densities for populations cultured at low density (r, $r \times r$) and populations cultured at high densities (K, rK, and $r \times r$K). The bars are standard errors. The derivation of the various lines is described in the text. The measurements for the r- and K-populations shown as solid histograms were made after eight generations of selection. Measurements for the other populations were made after 223 generations of selection in the r-environment (see fig. 6.22). The bold numbers are the fitnesses of the K-populations relative to the appropriate r-population.

The r- and K-populations are presumably genetically variable (the experiments with the rK and $r \times rK$ populations demonstrate this). Thus, the differences in growth rates measured in figure 6.23 reflect performance averaged over a number of genotypes. These growth rates can also be used to estimate mean fitness of the r- and K-populations. Let λ be the growth rate of the r-population at a particular density and λ_K be the same growth rate for the K-population. A generation in these serial transfer systems is about three weeks, so the relative fitness of the K-population can be estimated as $(\lambda_K/\lambda_r)^3$. These relative fitness values are shown in figure 6.23. At extreme densities, we see that the mean fitness differences were between 8 and 53 percent. Thus, fitness differences between individual genotypes were probably even greater than these values, indicating strong selection. Given this strong selection, it is not surprising that repeated selection experiments would yield similar results (since the effects of selection would overwhelm random forces such as drift).

All combined, these experiments provide compelling evidence that rates of population growth may respond to selection. Further, there are trade-offs involved in this evolution: Populations adapted to crowded conditions do more poorly at low density than do populations adapted to low-density conditions, and vice versa. These adaptations to high and low density have been dissected further. Larvae adapted to crowding show increased competitive ability for food, which is accomplished by increased feeding rates (see fig. 6.2; Joshi and Mueller, 1988; Mueller, 1988a; Guo et al., 1991; Santos et al., 1997). These larvae also differ in their two-dimensional foraging behavior: The K-populations are predominantly the rover phenotype, and the r-larvae are predominantly the sitter phenotype (Sokolowski et al., 1997). Larvae kept at high density also evolve increased pupation height, which reduces mortality among pupae (see fig. 6.3; Mueller and Sweet, 1986; Joshi and Mueller, 1993). Several earlier studies with *Drosophila* noted that populations newly introduced into the laboratory and kept near their carrying capacity show a gradual increase on the equilibrium population size (Ayala, 1965b, 1968; Buzzati-Traverso, 1955).

INCREASED RISK OF POPULATION EXTINCTION WITH INBREEDING

A practical application of information about population dynamics is to the prediction of the chance of population extinction (Ewens et al., 1987; Lande, 1993; Mangel and Tier, 1993, 1994; Ludwig, 1996, 1999). Models of population extinction have shown that an important component of this probability is the chance of rare catastrophic events and environmental variation. Typically we would expect laboratory studies to shed little light on these quantities since they are characteristics of specific environments and habitats. However, another important issue in conservation biology has been the effects of inbreeding on the risk of population extinction (Allendorf and Leary, 1986; Lande and Barrowclough, 1987; Caro and Laurenson, 1994). This issue is especially interesting since many endangered species exist as small populations that increases the likelihood of matings between close relatives.

With *Drosophila* we can simultaneously inbreed populations and determine the effects of inbreeding on population growth. In fact, this sort of experiment was carried out 20 years ago to address questions unrelated to conservation ecology: Mueller and Ayala (1981d) created 24 populations of *D. melanogaster*, each homozygous for a different second chromosome sampled from nature. Because the second chromosome is nearly 40 percent of the genome, these populations had probabilities of alleles being identical by descent (F) of nearly 40 percent. In addition to these 24 inbred populations, there was a single outbred population created by mass crossing of all 24 inbred populations on three separate occasions.

For each of these populations, the rates of population growth in the serial transfer system were estimated over a range of densities (see chapter 3 for additional discussion). We have used simple first-order difference equations to summarize the asymptotic rates of population growth at each density (Mueller and Ayala, 1981c). Although these rates of growth are approximations for the true serial transfer system, the goal of this analysis is to evaluate the relative differences between populations rather than achieve precise numerical estimates of population extinction.

To estimate probabilities of extinction, we used the Markov chain model developed by Ludwig (1996, 1999). This model assumes environmental variation that is normally distributed on a natural log scale of population size. To make the Markov chain finite, an upper bound on population size was specified at 150 percent of the carrying capacity. Because the population-growth models may predict sizes greater than this upper bound, the models are modified to prevent this. In the Markov chain, probabilities of going from size j to the maximum size in a single generation also include the probability of going from size j to all other population sizes greater than the maximum.

The results show that inbreeding leads to a pronounced increase in the chance of extinction (fig. 6.24). This is not unexpected, since a major conclusion from these studies was that inbreeding caused severe reductions in population growth rates at low densities, relative to the outbred population (Mueller and Ayala, 1981d). However, another observation from the original studies was that there was little difference between the outbred and inbred populations in rates of population growth at high densities. Nevertheless, we see that the outbred population still has a substantial, although slightly reduced advantage at high densities. (At the lowest starting density, there are only two inbred populations with lower probabilities of extinction, whereas at the higher densities there are six.) This result is certainly related

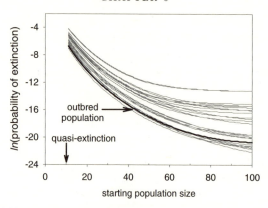

FIG. 6.24. Probability of population extinction after 1000 generations in 24 inbred populations (thin-gray lines) and one outbred population (thick-black line) as a function of the initial population size. Probabilities were computed from the Markov chain model of Ludwig (1999), assuming that the standard deviation of the environmental variation was 0.5 and that quasi-extinction occurred at 10 individuals. Growth for each population was modeled by the theta logistic equation; and parameter values for each line are given in Mueller and Ayala (1981c).

to the fact that even if the population starts at a high density, to go extinct it must pass through low densities, at which point the ability of the population to grow at low density is crucial. We conclude that for species that do not normally inbreed, inbreeding can have substantial negative effects on long-term population persistence.

This problem has also been studied directly by Bijlsma and his colleagues (Bijlsma, Bundgaard, et al., 1997; Bijlsma, Bundgaard, and Boerema, 2000). In these studies, populations of *D. melanogaster* were inbred to different levels. Replicate small populations were then followed over time and population extinction recorded. The results show a clear increase in rates of extinction for inbred populations. Additionally, environmental stresses such as high temperature or ethanol increased the chances of population extinction much more in inbred populations than in outbred populations.

232

EVOLUTION OF POPULATION STABILITY

The experimental research reviewed earlier in the chapter has shown that the stability of *Drosophila* populations may be manipulated by varying the levels of food supplied to larvae and adults. It is therefore possible to create replicate populations of *Drosophila* that live in two alternative environments, one conducive to stable population dynamics and the other not. These populations may then adapt to their respective environments, and evidence of altered population dynamics can be obtained. In particular, it is of great interest to determine if populations kept in an environment that causes instability may evolve life histories that yield stable population dynamics. We are especially interested in evolution by selection at the individual level (as opposed to some type of group selection mechanism). The end of chapter 2 included a review of several theoretical models for the evolution of population stability at the individual level. These models develop the plausibility of evolution increasing population stability, but their assumptions are untested.

The r- and K-populations described earlier are not the best material for these experiments, for a variety of reasons. These include (1) low effective population size in the r-populations relative to that of the K-populations lead to more rapid loss of genetic variation and fixation of deleterious mutations (Mueller, 1987); (2) Reproduction at different ages in the r- and K-populations lead to different levels of age-specific selection, which may be confounding; and (3) Simultaneous density effects on two life stages (larvae and adults) make it impossible to unambiguously assign selection at a specific life stage to particular types of evolution. To overcome these problems several new populations were created (fig. 6.25). The CU populations experience crowding only during the larval life stage (see fig. 6.25). The UUs serve as controls and are uncrowded during their larval life stage.

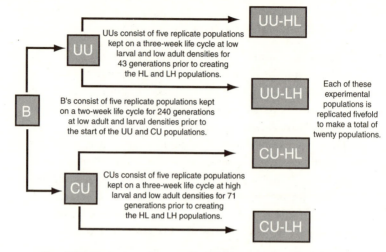

Fig. 6.25. Derivation of twenty *Drosophila* populations used to study the evolution of population stability. The B, UU, and CU populations are all maintained with breeding populations of 1000 to 2000 adults.

The UU and CU populations served as sources for experiments used to study the evolution of population stability. Samples from each CU and UU population were placed in two different environments: the LH environment, which tends to produce population cycles, and the HL environment, which tend to give rise to a stable-point equilibrium (see fig. 6.25). In the discussion that follows we refer to populations kept in the LH or HL environments as LH or HL populations respectively.

An important aspect of selection in the LH and HL populations is the potential for population cycles to cause occasional bottlenecks in the numbers of breeding adults. In figure 6.8, for instance, populations occasionally dip to 60 adults in the LH populations. Likewise, in Nicholson's blowfly experiments population size was reduced to fewer than 100 breeding adults for short periods. These bottlenecks can cause inbreeding depression, which typically reduces female fecundity (Marinkovic, 1967). Because popu-

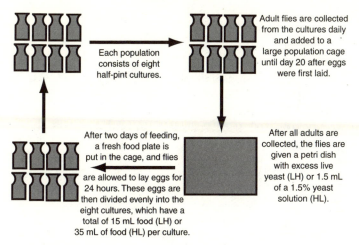

Each population consists of eight half-pint cultures.

Adult flies are collected from the cultures daily and added to a large population cage until day 20 after eggs were first laid.

After all adults are collected, the flies are given a petri dish with excess live yeast (LH) or 1.5 mL of a 1.5% yeast solution (HL).

After two days of feeding, a fresh food plate is put in the cage, and flies are allowed to lay eggs for 24 hours. These eggs are then divided evenly into the eight cultures, which have a total of 15 mL food (LH) or 35 mL of food (HL) per culture.

FIG. 6.26. Life cycle and maintenance of LH and HL *Drosophila* populations used to study the evolution of population stability (after Mueller et al., 2000).

lation stability is often a function of maximum rates of reproduction at low density, inbreeding may enhance population stability in a highly fecund species such as *Drosophila*.

Consequently, for these experiments the procedures used in figure 6.8 were altered so that populations would be sufficiently large that bottlenecks would not reduce total numbers below 1000 breeding adults. The procedures for maintaining these experimental populations maintained many of the features of the experimental system used in figure 6.8, but the total number of cultures to maintain each population was increased about eightfold (fig. 6.26). This produced adult population sizes that were typically between 2000 and 6000 individuals (fig. 6.27).

The first assessment of this experiment occurred after 45 generations of selection (Mueller et al., 2000). Those results along with an additional 23 generations are shown in figures 6.27–6.29. Mueller et al. (2000) found no evidence that the dynamics of the unstable LH populations about their equilibrium point had been altered due to 45 generations of

235

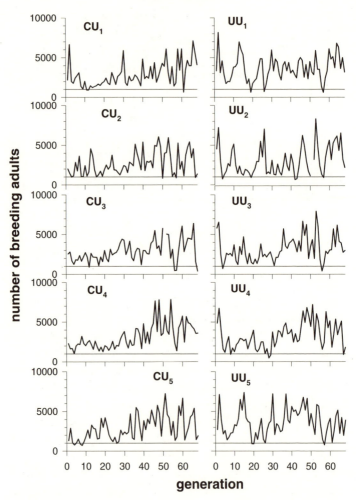

Fig. 6.27. Adult population sizes in the 10 HL populations of *Drosophila melanogaster* over 68 generations of maintenance by the techniques outlined in figure 6.26.

selection. However, there was strong evidence of evolution in these populations in response to density (Mueller et al., 2000; see also fig. 6.30). One interpretation of these results is that selection may in fact be taking place but was insufficient in magnitude to be detected by our methods.

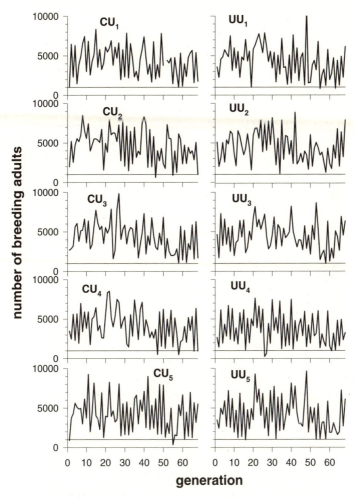

Fig. 6.28. Adult population sizes in the 10 LH populations of *Drosophila melanogaster* over 68 generations of maintenance by the techniques outlined in figure 6.26.

From figures 6.27 and 6.28, it is clear that the LH populations tended to be larger and that all populations went below 1000 adults only rarely in any single generation. Each generation, the total dry weight of the adults was recorded (fig. 6.29). These data demonstrated that the adults in the

237

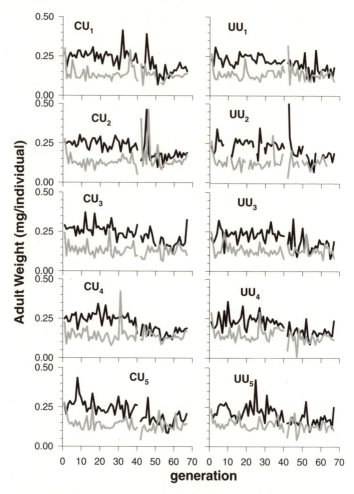

Fig. 6.29. The average weight per adult in the 10 LH (light-grey line) and 10 HL (black line) populations of *Drosophila melanogaster* over 68 generations of maintenance by the techniques outlined in figure 6.26. The HL populations are uniformly and significantly heavier than the LH flies, due to the much higher larval densities in the LH populations. The CU-HL populations show a significant decline in average size, which accompanies their increase in adult population over the same time period (see fig. 6.27).

LH populations were much smaller than adults in the HL populations (Mueller et al., 2000) due to the higher larval densities. In the novel HL environment, the CU populations showed a significant increase in numbers of adults over time (fig. 6.27) and a significant decline in mean size (see fig. 6.29; Mueller et al., 2000). These changes may reflect adaptation of the CU populations to the environments characterized by low larval density. These changes could come about from increased egg-to-adult viability that accompanied the declining feeding rates in these populations (fig. 6.30), since these traits appear to be genetically correlated (Borash et al., 1998; Borash et al., 2000).

We used a second-order RSM model to estimate the stability determining eigenvalue of each population:

$$\ln(N_{t+1}/N_t) = a_1 + a_2 N_t^{\theta} + a_3 N_t^{2\theta} + a_4 N_{t-1}^{\theta} \qquad (6.8)$$
$$+ a_5 N_t^{\theta} N_{t-1}^{\theta}$$

Results using this model showed no consistent trend in the 10 LH populations (table 6.7). In 6 out of 10 cases, the magnitude of the eigenvalue decreased. We also examined the autocorrelation function in the first and last 15 generations of the experiment. Presumably, evolution that affects population dynamics might result in a change in the magnitude or sign of the correlations over the course of the experiment; however, the autocorrelations give no suggestion of a consistent change between the start and end of the experiment (fig. 6.31). Our conclusions are similar to those of Mueller et al. (2000), who examined the first 45 generations: There is no evidence that the stability properties of the LH populations have changed, despite evidence of adaptation to the high and low larval crowding in these environments.

It is difficult to say precisely why there has been no evolution of population stability. It seems unlikely that genetic variation does not exist for important life history

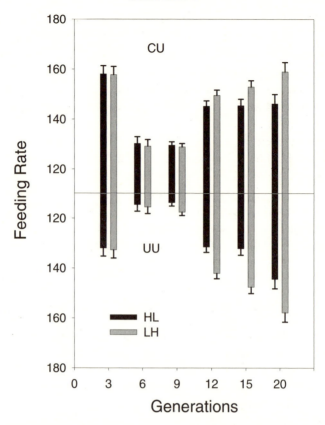

FIG. 6.30. The change in feeding rates of the CU-HL and CU-LH populations (top panel) and the UU-HL and UU-LH populations (bottom panel). Each bar is the mean of the five replicate populations, and error bars indicate 95% confidence intervals. Initially, there are no differences between HL and LH populations, although the CU populations feed faster than the UU populations due to their history of high larval densities. With time, the feeding rates of the LH populations exceed those of the HL populations due to the increased larval density in the LH cultures relative to the HL cultures.

characters in *Drosophila*, given the information we have already reviewed. However, it may be that the strength of selection on characters that would ultimately affect population stability is weak and thus difficult to observe even after 68 generations. It may also be that the characters that might

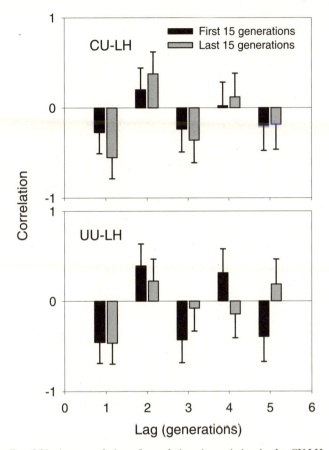

Fig. 6.31. Autocorrelation of population size variation in the CU-LH and UU-LH populations. Correlations were determined on the variation in the first 15 (1 to 15) and last 15 (54 to 68) generations. Any linear trends were removed from the series prior to the calculation of the autocorrelations. The error bars indicate 95% confidence intervals based on the five replicate populations.

evolve to change population stability in turn affect other fitness components negatively and thus do not evolve.

In any case, it is clear from this experiment that one class of environments can cause unstable dynamics. Over ecologically relevant time spans, populations of *Drosophila* do not

241

TABLE 6.7. Estimated stability-determining eigenvalue for the ten CU- and UU-LH populations during the first and last 15 generations of the experiment*

Population	First 15 (generation 1–15)	Last 15 (generation 54–68)
CU_1	−1.89	−0.78
CU_2	−0.70	−0.76
CU_3	−0.73	−0.95
CU_4	−1.08	−0.63
CU_5	−0.75	−1.53
UU_1	−1.43	−0.61
UU_2	0.42**	−0.80
UU_3	−0.92	−0.77
UU_4	−0.96	0.37**
UU_5	−1.12	−0.30

* In each case the second-order model (Equation 6.8) was used with θ set to 0.5.
** Complex eigenvalue.

seem to be capable of changing population stability by adaptation to these particular environments.

SUMMARY

- Models of the dynamics of *Drosophila* populations suggest that the relative levels of food given to larval and adult stages are crucial for the ultimate stability of the populations. The conditions that favor stability are high levels of food for larvae and low levels of food for adults. Cycles and other departures from stable-point equilibria are predicted when larvae are given low levels of food and adults are given high levels of food.
- The predictions from these models are supported by empirical research with replicated laboratory populations of *D. melanogaster*.
- Density-dependent natural selection results in adaptations that affect rates of population growth in *D. melano-*

gaster. An individual characteristic that increases in crowded environments is the larval feeding rate.

- When *Drosophila* are kept in environments that result in population cycles, evolution of traits (e.g., feeding rates) are observed, but the stability characteristics of the populations remain unchanged.

CHAPTER SEVEN

Natural Populations

This chapter marks a major departure from the others, since here we consider natural populations rather than model laboratory systems. Natural populations are, in many ways, the antithesis of model populations. In nature, environmental factors vary over time and space, sampling efforts may not be standardized, basic understanding of the role of density and age-specific effects on mortality and fertility may be lacking, and the impact of other species—both competitors and predators—may be unknown. (This is not a comprehensive list of the liabilities of natural populations.) Why, then, should we bother to study natural populations at all? Certainly, an appreciation for the types of dynamics observed in natural populations should motivate the questions addressed by research with model systems. If natural populations were rarely chaotic, a great expenditure of time and energy to uncover when model systems are chaotic would seem pointless. However, if natural populations were rarely chaotic and model systems were almost always chaotic, some reconciliation of these differences would be warranted. What is clear is that the factors responsible for the dynamic properties of natural populations may often be difficult to infer.

Our focus in this chapter is to review studies that span the range of techniques discussed so far. Our goal is to explore the strengths and weaknesses of these approaches, as applied to natural populations, rather than to be comprehensive in our review of natural systems. For instance, we do not review one of the best-documented cases of population cycling in nature, the lynx-hare system, since it is clearly a predator-prey cycle and has been reviewed several times previously

(Royama, 1992, chapters 5 and 6). Indeed, it may ultimately be the case that many if not most cycles in natural populations arise from interspecific interactions.

SIMPLE MODELS

The work of Hassell et al. (1976) was the first major effort to analyze data from many natural populations and infer their stability behavior. Stability was assessed by obtaining parameter estimates from the simple discrete-time model:

$$N_{t+1} = \lambda N_t (1 + a N_t)^{-\beta} \qquad (7.1)$$

The stability of this model depends on the values of β and λ (see equation 3.1 and the discussion that follows). The parameters a and β were estimated by regressing "observed mortality" on N_t. The "observed" mortality was in fact the quantity $\log[N_t \lambda / N_{t+1}]$. Thus, the observed quantities N_{t+1} and N_t were transformed by the quantity λ, which was estimated indirectly. The base value of λ was based on an estimate of maximum fertility, which was reduced further by a number of density-independent factors. In the case of the winter moth, for instance Hassell et al. (1976) reduced maximum fertility after taking into account (1) mortality between prepupal and adult stages, (2) mortality due to the parasitoid *Cyzenis albicans*, (3) mortality due to microsporidian disease, (4) mortality due to other larval parisitoids, and (5) mortality due to the pupal parasitoid *Cratichneumon culex*. These factors were significant and were responsible for reducing λ, in the case of the winter moth, from 100 to 5.5. The regressions that gave rise to β treat λ as a constant and therefore do not reflect its uncertainty.

Morris (1990) compared two additional techniques to the techniques used by Hassell et al. for estimating the parameters of equation 7.1. The first additional technique was similar to the method just described except that mortality was not log transformed. The second additional technique used the observed time series to estimate directly the values of $a, \beta,$ and λ. Interestingly, Morris's estimates of β and λ showed much less precision when estimated from the time series. We suggest that the high precision obtained by Morris for his estimates of λ and β by the first two techniques is illusory and a consequence of ignoring the variability in λ as described previously. Morris noted that the parameter estimates and their confidence intervals were sensitive to the method of estimation. More pointedly, we feel that this difference arises by treating uncertain parameters (λ) as known constants.

Although the approach of Hassell et al. was useful for framing the problem of population stability, their techniques suffer from several other problems. Certainly, there is no need to restrict these analyses to equation 7.1. It is also dangerous to restrict the analysis of time-series data to first-order equations for the reasons outlined in chapter 2. In all fairness, Hassell et al. were extremely cautious about the interpreting their results. The techniques developed by Turchin and Taylor (1992, reviewed next) relax many of these assumptions and thus ought to be more robust.

The major conclusion by Hassell et al. was that few populations showed cyclic or chaotic dynamics. Those that did were unusual populations such as agricultural pests (Colorado potato beetle) or laboratory populations (blowflies). The application of more robust techniques to natural populations has not completely reversed this view but has nudged more populations into the cyclic and chaotic regions (Turchin and Taylor, 1992). Nevertheless, the predominant impression left by Hassell et al. stills remains; that most natural populations appear to fall within the deterministically stable region of dynamic space.

SURVEYS USING RESPONSE SURFACE METHODS

The studies we review here were pioneered by Turchin and Taylor (1992) and were later extended by Ellner and

Turchin (1995). Even within the constraints of working with natural populations, there are some studies that yield data that are relatively more amenable to rigorous analyses. Desirable features of samples from natural populations include the following: (1) Samples are taken at regular time intervals in the same location. If samples are taken annually, there is no need to worry about correlations between samples that arise due to seasonal variation. If multiple samples per year are taken, then seasonal variation is a potential problem. (2) The effort and techniques used for collecting census information should be standardized and constant across time intervals. Unfortunately it is often difficult to determine the quality of census records for natural populations merely by inspection. If we were looking at counts of eggs laid by a single female *Drosophila* in a day, for instance, numbers above 200 would automatically signal an error, since this is far above what has been observed for fruit flies. Except for negative numbers, there is almost no set of observed population counts whose numerical value would similarly inform us of erroneous experimental technique. This makes the evaluation of historical data and published records problematic.

Turchin and Taylor (1992) used time-series analysis and the response surface method (RSM) technique to infer the behavior of natural populations from their census data; Ellner and Turchin (1995), however, looked for indications of chaotic versus nonchaotic dynamics by estimating Lyapunov exponents for each population using the RSM technique and other regression models. The results of these studies are summarized in table 7.1. Although a greater proportion of the species in table 7.1 show cycles or chaos compared with those studied by Hassell et al. (1976), such species are still in a minority. Earlier analysis of data on the incidence of measles in humans had suggested chaotic dynamics (Sugihara and May, 1990), a conclusion supported initially by the analysis of Ellner and Turchin (1995).

247

TABLE 7.1. Summary of the dynamic behavior of natural populations studied by Turchin and Taylor (1992) and Ellner and Turchin (1995)

Dynamic Behavior	Population/Species
Chaos	*Phyllaphis fagi*
Quasiperiodicity	*Lymantria dispar, Zeiraphera diniana,* lynx, Belyak hare
Stable cycles	*Drepanosiphum platanoides*
Stable equilibrium (oscillatory approach)	*Hyoicus pinastri, Dendroctonus frontalis, Lymantria monacha, Bupalus piniarius, Hyphantria cunea, Vespula spp.,* Arctic fox, colored fox
Stable equilibrium (exponential approach)	*Choristoneura fumiferana, Dendrolimus pini, Panolis flammea*
Not chaotic	Red grouse*, wolverine, martin, muskrat, red and arctic foxes, partridge, rabbit, snowshoe hare*, weasel, *Ceroplastes floridensis, Parlatoria camelliae, Trips imaginis,* measles

*One of three tests suggest chaos.

However, when Ellner and Turchin reanalyzed these data, accounting explicitly for seasonal variation, they obtained a negative Lyapunov exponent, suggesting mild stability rather than chaos. This example underscores the importance of taking into account seasonal contributions to observed patterns of variation when data are collected more frequently than once a year.

Unlike Turchin and colleagues, who relied on nonlinear regression techniques for providing estimates of model parameters, Dennis and Taper (1994) used maximum-likelihood techniques. Although maximum-likelihood estimates have many desirable properties, their application requires knowledge of the statistical distribution of the random noise, a constraint that does not apply in the case of regression. Dennis and Taper assume that on a logarithmic scale, errors are distributed normally with a common variance independent of population density. Although this sounds reasonable, there are few data that support these assumptions. Ideally, to test the assumption of a common variance across densities, one should collect independent replicated observations of population growth at a range of population densities. Whereas this type of data would be difficult if not impossible to collect in natural populations, it has been collected in laboratory populations (fig. 7.1). The data in figure 7.1 show little effect of density on the variance (on a log scale) of population growth rates over 23 genetically different populations of *Drosophila melanogaster* (Mueller and Ayala, 1981d).

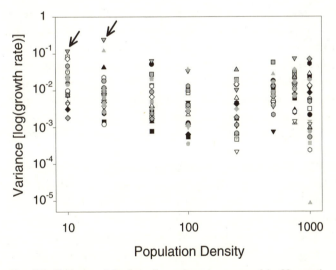

FIG. 7.1. Variance of the log of growth rates measured in 23 genetically different populations of *Drosophila melanogaster*. Each population was homozygous for a different second chromosome sampled from nature and is represented by a different symbol. There are occasional examples of genotypes that affect the variance but no consistent differences by density.

249

CHAPTER 7

DETAILED STUDIES OF SINGLE POPULATIONS

In this section, we review a few studies of natural populations that focus on examining the dynamics of one or a few
species for a prolonged period of time. These studies yield a
relatively detailed understanding of the basic biology of the
species of interest, an aspect that makes it possible to appreciate better some of the details of the population dynamics
observed.

Soay Sheep and Red Deer

We first discuss a multiyear study on Soay sheep (*Ovis
aries*) and red deer (*Cervus elaphus*) that has successfully
yielded a sophisticated understanding of population regulation (Clutton-Brock et al., 1997). In this study, variation in
population size for Soay sheep and red deer was recorded
for about 9 and 20 years, respectively. Although several age
classes were followed in each population, the fundamental
difference between the two species is evident from data on
total numbers (fig. 7.2). The red deer population shows relatively stable dynamics, with small changes from one year to
the next, whereas the Soay sheep population shows dramatic
oscillations from year to year.

The effects of crowding on survival and fertility are well
documented for these two ungulates (Clutton-Brock et al.,
1997). This detailed knowledge is ultimately important to
understanding why Soay sheep do not settle down to an
equilibrium as do the red deer. Soay sheep, unlike red deer,
are capable of producing large numbers of offspring in any
year, and can easily exceed the numbers that can be supported by the environment, giving rise to large population
crashes. The key life-history characteristics that contribute
to the population dynamics differences between these two
species are as follows: (1) Female Soay sheep have offspring
in the first year of life, whereas it takes red deer three to four
years before females reproduce. Moreover, Soay sheep may

250

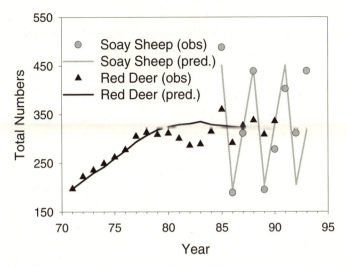

FIG. 7.2. Total population census of Soay sheep on the island of Hirta, St. Kilda, and red deer in the North Block of Rum, Inner Hebrides. Symbols show the observed census counts, and lines are predicted sizes from an age-structured model (after Clutton-Brock et al., 1997).

have twins, and red deer almost always give birth to a single calf. (2) Soay sheep are more likely than red deer to give birth in consecutive years. Females of both species become pregnant in the fall and give birth in the spring. However, young Soay sheep, become independent of their mothers early in the summer, permitting the females to feed and put on enough additional weight to support another pregnancy in the fall. Red deer calves continue to feed from their lactating mothers throughout the summer, making it nearly impossible for a single female to have calves in two consecutive years. (3) Overwinter survival drops precipitously with total density for Soay sheep but only gradually for red deer.

Ultimately, though, there is no replication of populations in this study. The extent to which we believe that the observed dynamics are driven by the underlying life-history differences just enumerated rests on the logic of those arguments, not on our ability to manipulate these factors and repeatedly see the predicted behavior. Nevertheless, the strength of this system is the ability to carefully document survival and reproduction among individuals in the population. This permits a more sophisticated understanding of the population-level phenomena by extrapolation from the behavior of individuals.

Clutton-Brock et al. (1997) also suggested that feral populations may be more likely than wild populations to exhibit cycles, owing to artificial selection for high fertility that occurred when the ancestors of the feral animals were in captivity. They add that feral populations are often established in areas free from predation, potentially exacerbating this effect. The validity of this generalization may, however, vary across species. Moreover, one would expect that if domesticated animals released in the wild came back into contact with wild populations, the domesticated traits would not persist in subsequent generations. After all, fertility is always under strong selection in natural populations. Thus, if fertility in natural populations is below what can be achieved by artificial selection, it is likely due to trade-offs in other fitness components. However, if feral populations are protected from introgression from natural populations, the conjecture of Clutton-Brock et al. (1997) may be valid, at least for some species.

Perennial Grass, Agrostis scabra

Tilman and Wedin (1991) studied monocultures of the perennial grass *Agrostis scabra*, maintaining different levels of soil nitrogen and recording live biomass and litter (dead biomass) yearly. They found that treatments with the highest nitrogen levels exhibited large and erratic fluctuations in live biomass from year to year that could not be explained by environmental variation. The biology of this system suggests that growth of new biomass in the current year is a function of the litter left over from the previous year's growth. A very dense litter inhibits growth by intercepting light, and decay of the litter removes this inhibitory effect.

Tilman and Wedin's model of the dynamics of live biomass and litter biomass suggested that, at very high nitrogen levels, live biomass could behave chaotically over time. The parameters of the model were estimated by experimentally varying nitrogen levels and then measuring the biomass produced from those plots. The time series were, however, too short to be used in estimating the model parameters.

In this system, there is a time delay of one year between the production of large amounts of litter and the density-dependent reduction of plant growth. This delay can ultimately drive the system to cycles or chaos when productivity is very high (e.g., at high soil nitrogen levels). Even though the plants were grown outdoors, conditions were only seminatural because experimenters controlled seed density, water, and nitrogen. Indeed, the strength of this study lies in the experimenter's ability to manipulate aspects of the environment that are considered important for determining stability.

Lemmings and Voles

Cycles in the dynamics of populations of small mammals have interested ecologists for more than 70 years (Elton, 1924). A number of well-studied natural populations of small mammals in Northern Europe have been sampled at regular intervals for extended periods; we show some of these data in figure 7.3. Many explanations for these cycles have been put forward and discussed in the literature (review by Batzli, 1992), so we do not review them here. It is clear that vole dynamics are affected by predator species as well as intraspecific competition. However, there is much less information about the dynamics of these predator populations, and inferences about the dynamics of vole populations are typically made from single-population time series alone.

The techniques of time-series analysis and nonlinear RSM techniques have been applied to these data by Turchin (Turchin, 1993, 1995b) and Falck et al. (1995a, 1995b).

253

Turchin's results suggested that the more northern populations (Norway, Russia) were chaotic, whereas those further south (below about 60° latitude) were not. Two northern and one additional southern population are shown in figure 7.3 along with their estimated Lyapunov exponents. These conclusions have been questioned by Falck et al. (1995a, 1995b). We now review this critique since

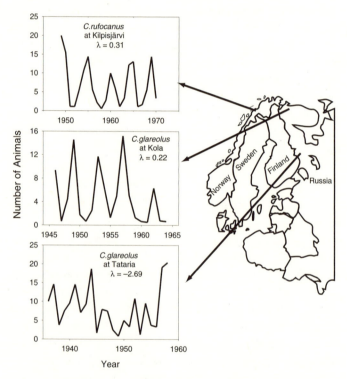

FIG. 7.3. Variation in population size for voles of the genus *Clethrionomys*. All three populations are from Fennoscandia, although the population in the bottom panel is from a site farther south than those shown in the top two panels. λ's are the estimated Lyapunov exponent (after Falck et al., 1995b), and their values indicate that the populations shown in the top panels may exhibit chaotic dynamics. Note the different axes among the three panels.

it raises some general problems about the RSM techniques and ecological sampling.

A major criticism by Falck et al. (1995a, 1995b) was that the estimated Lyapunov exponents were not accompanied by estimates of uncertainty in the form of confidence intervals; this problem has been acknowledged by Turchin (1995b). Falck et al. used the bootstrap to simulate new time series from the observed residuals. If the observed time series is $N_1, N_2, \ldots N_m$ and the nonlinear function $f(N_{t-1}, N_{t-2}, \ldots, N_{t-k})$, then associated with each observation is a residual:

$$\varepsilon_t = N_t - f(N_{t-1}, N_{t-2}, \ldots, N_{t-k})$$

The residuals are then sampled with replacement to generate a new time series. The Lyapunov exponent is estimated from the simulated time series and saved. This procedure is repeated many times, and the replicate Lyapunov exponents can be used to create a confidence interval. As pointed out by Efron and Tibshirani (1993), this technique is only as good as the regression model because ε_t includes observational errors, environmental noise, and lack-of-fit errors. If this last term is large, it will inflate the size of the confidence interval. In a later paper, Falck et al. (1995b) used a different procedure for generating bootstrap samples that sampled neighboring observations directly. The confidence intervals generated from neighboring observations were similar in size to the intervals generated from the residuals. Thus, their results suggest that estimated Lyapunov exponents tend to be biased: Positive Lyapunov exponents tend to be too small, and negative exponents tend to be too large. This pattern must be taken into account both when estimating the Lyapunov exponent and when estimating its confidence interval.

Falck et al. (1995b) also note that the confidence interval around many of the positive Lyapunov exponents estimated for the vole data includes zero. Therefore, they believe that the evidence overall does not strongly support chaotic dynamics for voles. On the other hand, Turchin (1995b) points to the consistent appearance of positive Lyapunov exponents in independent populations when he suggests there is strong evidence for chaotic dynamics. Turchin's point is well taken, although the argument ultimately hinges on the extent to which we can consider multiple natural populations of the same species to be replicates. In the strict sense, such sets of populations are not replicates, since there is no way we can ensure that each population experiences the same conditions such as environment and predation pressures. The implicit argument here is, of course, that the major environmental variables affecting dynamics are roughly divided along a north-south gradient. Thus, any populations above 60 degrees latitude will experience a "common" environment and can therefore be treated as replicates of a common environmental regime. One previously held notion of the north-south dichotomy was that generalist predators are more important south of about 60 degrees, and that they have a generally stabilizing effect on vole dynamics. Turchin's position is that the combined forces of intraspecific competition and predation from generalist and specialist species ultimately make any vole population north of 60 degrees grow chaotically.

It appears that the differences between the approaches of Turchin (1995b) and Falck et al. (1995b) finally boil down to an argument over the power of replication. Turchin is arguing that the repeated observation of the same finding bolsters one's confidence that there is a general set of forces that shapes the dynamics of a set of populations. Falck et al. (1995b) seem to argue against the benefits of replication in pointing out that it may be preferable to have a single time series of 200 data points rather than 10 time series consisting of 20 points. They fail to appreciate that even infinite knowledge of a single population does not permit us to make any generalizations about population processes.

In an attempt to solidify the general understanding of the vole system, Turchin and Hanski (1997) have developed a theory to incorporate the effects of predation and a seasonal environment. Their model assumes that there are specialist predators (e.g., weasels) and generalist predators (e.g., foxes, badgers, and feral cats). Moreover, the intrinsic rate of increase of the voles and the specialist predators varies as a sine wave to reflect seasonality. Under this model, the north-south dichotomy arises because the generalist predators, which have a stabilizing effect on vole dynamics, are more common in the south.

Turchin and Hanski (1997) estimate some of their model parameters from the natural populations and then use the models to predict dynamics. Although the model effectively predicts the north-south dichotomy in vole dynamics, more empirical information is needed on the various predator populations to confirm the basic correctness of the model.

Red Grouse

Red grouse (*Lagopus lagopus scoticus*) are popular game birds in England and Scotland. The numbers of birds caught by hunters on different estates have been catalogued for many years. Some of these data have made their way to biologists who were taken by the apparent cycles that appear in these records (Middleton, 1934; MacKenzie, 1952). Time-series analysis of these records demonstrate the presence of true cycles in some populations (Williams, 1985). The flavor of these data can be obtained from one data set from northern Scotland (fig. 7.4).

Moss et al. (1996) review the evidence against several hypotheses that have been proposed for these cycles, including predator-prey interactions and high parasite burdens. To investigate the importance of the number of breeding males, they experimentally removed males from one population of red grouse and compared the population-size variation with that of an unmanipulated control population.

257

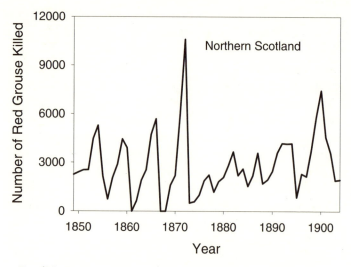

Fig. 7.4. Number of red grouse killed by hunters on an estate in Aberdeenshire, northern Scotland (data from Middleton, 1934).

This experiment clearly showed that male removal prevents the population from cycling. The areas in which males were removed also exhibited a reduction in female numbers. Male red grouse hold territories, and females must pair with a territorial male to mate. Although territorial males are sometimes found without female mates, the reverse does not occur. Apparently, the removal of males from a population also results in females leaving the population. The precise mechanism that affects the breeding success is not known but probably involves the number and age of the breeding population.

A recent study has provided strong support for the idea that the red grouse cycles are a consequence of parasite burdens (Hudson et al., 1999). This theory suggests that during high levels of infection by the parasitic nematode *Trichostrongylus tenuis*, female fertility declines dramatically and results in population crashes. To test this, Hudson et al. treated replicate populations with anthelmintic drugs either once or twice in a 9-year period and compared population numbers over time with a control population that received no drugs. The cycles were nearly eliminated in the drug-treated populations, suggesting that the removal of this second trophic level was sufficient to stabilize the red grouse population. Whether this study can be used to conclude that parasites are the only cause of the cycles is a subject of debate (Hudson et al., 1999; Lambin et al., 1999). Nevertheless, it is clear that parasites play an important role in these cycles.

WHY IS CHAOS RARE IN NATURAL POPULATIONS?

The data gathered by ecologists so far suggest that only few natural and laboratory populations might have chaotic dynamics. Even in the laboratory, *Tribolium* populations are chaotic only with constant manipulation of the population by humans. Yet, many models of population dynamics admit the possibility of chaotic behavior. Why, then, do we not see chaos more often? This is a question to which the answer is not clear at present. It is possible, as some workers have suggested (Turelli and Petry, 1980; Mueller and Ayala, 1981b; Doebeli and Koella, 1995), that life-history evolution through natural selection on individuals in populations typically results in the evolution of demographic parameters to values that do not produce chaotic dynamics. The preliminary evidence from *Drosophila* suggests that if such evolution takes place, it is not rapid. However, the information we have derived also suggests that population dynamics in this species may vary widely among populations adapted to different density conditions. In nature, this could translate into wide variation of dynamics over space. Indeed, the evidence from voles and lemmings also supports this conclusion. It is also known that isolated chaotic populations do have greater chances of going extinct than do relatively more stable populations of similar mean size.

Consequently, environments that produce chaotic dynamics for a given species may seldom have viable populations. Thus, as a matter of sampling, we simply do not see many chaotic populations. These two explanations implicitly invoke different causes for chaotic dynamics, however. In the former case, the cause for chaos is assumed to be the genetically determined values of demographic parameters in the population; whereas in the latter case, chaos is assumed to be due to environmentally determined values of demographic parameters. This distinction is not always made explicitly but needs to be kept in mind whenever one is discussing the evolution of population dynamic behaviors.

Another possibility is that there are other biological details of populations that reduce the likelihood of chaotic dynamics. Several authors (McCallum, 1992; Rohani and Miramontes, 1995; Ruxton and Rohani, 1998) have suggested that population floors may be responsible for inhibiting chaos in natural populations. A population floor is simply a portion of the population that is invulnerable to density regulation. We may represent this as

$$N_t = f(N_{t-1}, \ldots, N_{t-k}) + \psi,$$

where ψ is the invulnerable fraction of the population. One could imagine ψ as representing a spatial refuge from competition or predation, or a relatively constant source of immigrants from large external populations. Rohani and Miramontes (1995) and Ruxton and Rohani (1998) add population floors to a variety of single-population growth models and host-parasitoid models. Although adding these floors had the general effect of making chaos more difficult to reach, there were interesting qualifications. Chaos reached by quasi-periodicity, for instance, was more resistant in these models to the effects of population floors than was chaos reached by period doubling.

At present, our current knowledge of theory and the biology of populations does not rule out the possibility of populations showing chaotic dynamics. Indeed, although several candidate natural populations have been studied, only a small minority of these actually seems to exhibit chaotic behavior. It may be that finding populations with the right life histories, environments, and historical accidents to reveal chaos is a rare event.

CHAPTER EIGHT

Conclusions

A HEURISTIC FRAMEWORK FOR VIEWING POPULATION
DYNAMICS AND STABILITY

In both theoretical and empirical studies in population ecology, two broad categories of models have been used extensively. At one extreme, there are very simple models, such as the linear and exponential logistic models (see chapter 2), in which a single difference or differential equation represents the recursion of adult numbers from one generation to the next. All details of the life history and ecology of the species are subsumed into a single expression that embodies the dependence of adult numbers in one generation on the adult numbers in the preceding generation through a humped functional form. Although some of these models may do a reasonable job of capturing gross features of the dynamics of certain laboratory populations, they ignore many aspects of the biology of the organism that are known to play a major role in determining vital rates—and, through them, the dynamics of the population. These models typically ignore both stage structure (i.e., the division of the life cycle into discrete, ecologically distinct stages such as larvae, pupae, and adults) and age structure within a given life stage. Thus, they cannot distinguish among life stages that vary in terms of how strongly their numbers correlate to density-dependent regulatory mechanisms. Similarly, no distinction can be made between different regulatory mechanisms that exercise their effect by affecting the densities of different life stages. These distinctions between which life stages are the triggers or targets of density regulation, as well as the magnitude of ontogenetic delays between the trigger and target life stages, can have profound effects on the stability of the ensuing dynamics (Gurney and Nisbet, 1985; McNair, 1989, 1995).

By trigger life stage, we mean the life stage whose density is the stimulus for a density-dependent phenomenon that plays a role in population regulation; by target life stage, we mean the life stage whose numbers are affected primarily by the operation of a particular density-dependent regulatory phenomenon. The ontogenetic delay between trigger and target life stages refers to the time lag between the triggering of a density-dependent regulatory mechanism and its ultimate impact on the number of individuals in the target life stage. This time lag depends on when in the course of an organism's life cycle the target and trigger life stages occur. In *Drosophila*, for example, density-dependent fecundity plays a regulatory role; fecundity responds primarily to adult density, and the adult stage is therefore the trigger life stage. The impact of this density-dependent regulation, however, does not fall primarily on the adult stage. The target life stage for density-dependent fecundity is the egg stage, whose numbers are affected primarily by the operation of this density-dependent regulatory phenomenon. The same regulatory mechanism can have multiple trigger life stages, but will typically have only one target life stage.

At the other extreme of the spectrum of population dynamics models are detailed species-specific models that explicitly incorporate many of the relevant details of the species' life history and ecology. This was the case with the larva-pupa-adult (LPA) model for *Tribolium* (Dennis et al., 1995) and the *Drosophila* model of Mueller (1988b) discussed earlier. As we have seen, these models provide a good understanding of the factors affecting the population dynamics of a particular species, and their predictions have successfully withstood careful and rigorous empirical testing. At the same time, though, these very detailed and species-specific models do not have much heuristic value. The simple models, on the other hand, have been of heuristic value in terms of understanding how simple considerations of density dependence can give rise to complex dynamics, especially in the presence of time delays between the triggering of density-dependent regulation and its actual impact on population density (e.g., May and Oster, 1976). Beyond this, however, the simple models also have limited heuristic value. In particular, they have a limited ability to provide experimenters with a conceptual framework for deciding which aspects of the organism's ecology and life history are likely to play a major role in determining population dynamics, and whether that role is likely to be stabilizing or destabilizing. In this section, we will build on ideas outlined by McNair (1995) to construct a heuristic framework for viewing the impact of a species' life history and ecology on the dynamics of its populations. We hope that this framework is especially useful for experimenters who need to be able to separate those aspects of the biology that are relevant to population dynamics from those that have, at best, a small role to play—and who are, in many cases, not enamored of elaborate mathematical formulations that often seem to have little bearing on the biology of any real organism.

Using a stage-structured model (eggs, larvae, pupae, adults) with larval food supply being the critical factor limiting population growth, Gurney and Nisbet (1985) showed that density regulation triggered by the number of larvae gives rise to cyclic fluctuations in adult numbers. The periodicity of these fluctuations is affected strongly by whether larval density feed backs on itself or has a regulatory influence on other life stages. McNair (1995) further elaborated on the importance of the relative positioning in the life cycle of those life stages that act as the trigger and target of density-dependent regulatory mechanisms. He used a stage-structured model with overlapping generations and within-stage age structure. In this model, adult density was fixed as the trigger for density-dependent regulatory feedback. The target life stages for feedback were then varied systematically across the ontogeny, and the dynamic behavior of the model was studied. The results showed that the local stability of the equilibrium numbers of adults as well as early and late-stage juveniles depended critically on the interplay between two factors: (1) the sensitivity of fecundity and stage-specific mortality to adult density and (2) the life stage affected by adult density–dependent mortality. In this model, a unique stable equilibrium could always be obtained for adult and juvenile numbers by choosing appropriate values for fecundity and mortality; this equilibrium was not affected by the target life stage of the mortality, the age dependence or independence of fecundity, or the distribution of maturation times in the juvenile stage. Once such an equilibrium was obtained, McNair examined how the strength of density-dependent fecundity and mortality affected stability. He found that a local equilibrium could always be destabilized by increasing the sensitivity of fecundity to adult density. This gave rise to sustained oscillations in adult and juvenile numbers, regardless of the target life stage of density-dependent mortality. However, when the sensitivity of fecundity to adult density was held at a level that ensured stability of the equilibrium, and the sensitivity of mortality to adult density was increased, the effect on the dynamics depended on the target life stage. Extremely sensitive adult density–dependent mortality of early juveniles destabilized the equilibrium, giving rise to sustained cycles. In contrast adult density–dependent mortality of late juveniles or adults was not destabilizing even if sensitivity to adult density was high.

McNair (1995) explained these results on the basis of an ontogenetic time delay between the triggering of density regulation and its subsequent feedback on adult density when the target of the regulation was the egg or early juvenile stage. Thus, adult density–dependent fecundity or early juvenile mortality affects the numbers of eggs and young juveniles, respectively. It takes time—specifically, the maturation time of juveniles—for these effects to have an impact on adult numbers, which act as the trigger for density-dependent regulation. As a result, there exists the possibility of population cycles if the density dependence is strong. The effects of the density-dependent regulatory mechanism, thus, must trickle up the ontogeny before they can exert feedback on the trigger life stage.

Effects of adult density on either itself or late juvenile density, on the other hand, trickle up to the triggering life stage much faster, leading to relatively stable dynamics. Neither ontogenetic delays inherent in the life history (e.g., the delay caused by maturation time of juveniles) nor the mechanism of density dependence are, in themselves, the causes of instability in population dynamics. Rather, it is the ontogenetic time lag between the life stages, acting as the trigger and target of density-dependent regulation, that determines the nature of the dynamics. For example, in a model similar to McNair's in many ways, Rorres (1979) showed that strong density dependence of fecundity is destabilizing if adults are the trigger life stage, and is stabilizing if juveniles are the trigger life stage. Obviously, if juvenile density triggers regulation of adult fecundity, this effect will trickle up to the trigger (juvenile) stage much faster than if adults are the trigger life stage.

In addition to the identification of the trigger and target life stages of a species and their relative position in the life cycle, there are other important aspects of the biology of species that have an impact on population stability. In addition to any time delays between the effect of a density-dependent regulatory mechanism on the target life stage and the trickling up of that effect to the trigger life stage, it is necessary to consider possible time delays between the triggering of a regulatory mechanism and its effect on the target stage. If generations are fully discrete, for example, adult density cannot possibly act in a regulatory manner by feeding back on juvenile mortality (fig. 8.1); whereas this is easily possible when generations are overlapping, and may

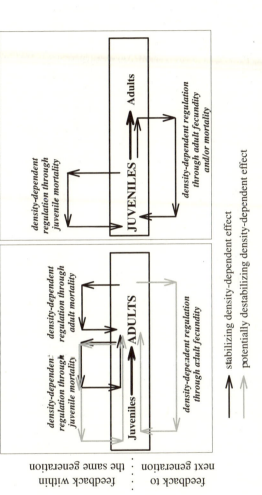

(A) Adult density triggers regulation   **(B) Juvenile density triggers regulation**

density-dependent regulation through juvenile mortality

density-dependent regulation through adult mortality

Juveniles ➤ ADULTS

density-dependent regulation through adult fecundity

density-dependent regulation through juvenile mortality

JUVENILES ➤ Adults

density-dependent regulation through adult fecundity and/or mortality

feedback to next generation ⋮ feedback within the same generation

➤ stabilizing density-dependent effect

➤ potentially destabilizing density-dependent effect

FIG. 8.1. Schematic representation of how stability of a population can be affected by the interplay between ontogeny and various density-dependent feedback loops (thin arrows). Model is for a species in which cohorts are segregated spatially (thick black vertical line separating the juvenile and adult stages), through either a fully discrete generation life cycles, or some other means. The thick black arrows represent ontogenetic transitions. The effect of juvenile density on adult mortality or fecundity in this case cannot be direct because of the segregation of life stages. It must, therefore, be mediated through physiological effects on adults of the density they experienced during the juvenile stage. Such an indirect effect of juvenile density on adult mortality or fecundity involves a long ontogenetic delay.

even be destabilizing if the mortality affects very early–stage juveniles (fig. 8.2).

Similarly, if generations are discrete and the trigger life stage is the juveniles, the effect of juvenile density on fecundity must be indirect, through the size of adults being reduced if they experienced relatively high densities as juveniles (see fig. 8.1). Thus, although the trigger and target life stages here are the same, juvenile density cannot feed back rapidly on itself due to the discrete generations. The effect of juvenile density in any generation will be felt only on the number of juveniles in the subsequent generation; this is a good example of what we mean by an "ontogenetic time delay." This effect can be destabilizing because the ontogenetic time delay is relatively long. On the other hand, if generations are overlapping, juvenile density can generate a stabilizing regulatory feedback loop through direct inhibitory effects on adult fecundity, which then feed back rapidly onto the juvenile stage (see fig. 8.2).

The distinction between organisms with discrete versus overlapping generations clearly exemplifies the possibility of a time delay between the triggering of a density-dependent regulatory mechanism and its actual effect on the target stage. The underlying issue here, however, is of the separation of cohorts in space and time. The importance of spatial separation of cohorts, although in a somewhat different context, has been stressed by Rodriguez (1998). Rodriguez studied a system modeled by differential equations, with time delays in density dependence built in, and found that such delays are destabilizing only when cohorts overlap in space. In a typically fully discrete generation laboratory system, however, cohorts do not coexist in either space or time. Therefore, the number of possible regulatory loops is reduced compared with the number in a system in which generations overlap in both time and space (contrast the number of such loops in figs. 8.1 and 8.2).

Systems with overlapping generations are often treated, at least for modeling purposes, as discrete-generation systems;

this is done by choosing appropriate time units for modeling and census, as in the case of the LPA model of *Tribolium* dynamics and experiments designed to test it (Costantino et al., 1995, 1997; Dennis et al., 1995; Benoît et al., 1998). Yet, in terms of what density-dependent regulatory loops are possible, the system may be far from approximating a truly discrete generation system, largely because of a lack of segregation of different life stages in space and time. Thus, adult beetles were able to cannibalize pupae even in the discretized *Tribolium* systems used by Dennis et al. (1995) and Costantino et al. (1995, 1997), whereas this would be impossible in a truly discrete generation system. Within the same kind of discretized system, however, when pupae were spatially segregated from the adults by providing them with a refuge, the stabilizing effect of pupal cannibalism by adults vanished, and adult numbers grew exponentially (Benoît et al., 1998).

There is also a difference in the possible range of effects that fecundity and mortality—the two processes through which density-dependent regulation operates—can have on the density of the target life stage. Fecundity basically causes recruitment into the first (youngest) juvenile stage and can therefore cause numbers to increase, whereas mortality affects recruitment into either later juvenile stages or the adult stage, and can only cause decreases in numbers. The one exception to this is neonate or egg mortality (e.g., egg cannibalism in *Tribolium*), which for all practical purposes implies a reduction in fecundity (broadly taken here to mean the level of recruitment into the first juvenile stage). Although both fecundity and mortality may slow population growth through density dependence, the maximum rates of population growth is determined by fecundity alone.

This point can be illuminated by an analogy. If a person is trying to drive a car at a constant speed imposed by the environment (i.e., a posted speed limit), the speedometer conveys information regarding the speed at any point

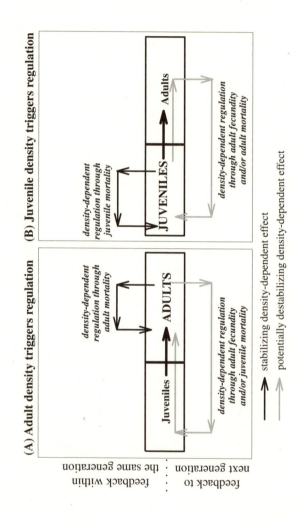

(A) Adult density triggers regulation

(B) Juvenile density triggers regulation

density-dependent regulation through adult mortality

ADULTS

Juveniles

density-dependent regulation through adult fecundity and/or juvenile mortality

density-dependent regulation through juvenile mortality

JUVENILES

Adults

density-dependent regulation through adult fecundity and/or adult mortality

⟶ stabilizing density-dependent effect

⟶ potentially destabilizing density-dependent effect

feedback within : the same generation

feedback to next generation : the same generation

in time, and the driver can, in principle, regulate speed by a combination of using the accelerator and the brake. However, the roles of the accelerator and brake are slightly different, even though both are components of the speed-regulating process. The accelerator can be used to either increase or decrease speed, whereas in the absence of acceleration, friction will tend to reduce speed. The brake, on the other hand, can only decrease speed. The accelerator, thus, is analogous to fecundity, with mortality playing the role of friction, whereas the brake is analogous to mortality. Now imagine a situation in which there is no speed-dependent control on the accelerator, which is fixed at a certain position. The driver, then, has to regulate speed based entirely on the brake. In this situation, it is easier to drive at a constant speed if the acceleration is fixed at a relatively low level. The analogy here is with systems in which mortality is density dependent and fecundity is density-independent, but the former is relatively high and the latter is low. Similarly, if the braking intensity is to be fixed at a speed-independent level, maintaining a constant speed will be easiest if base-

---

FIG. 8.2. Schematic representation of how stability of a population can be affected by the interplay between ontogeny and various density-dependent feedback loops (thin arrows), in a species in which generations overlap and cohorts are not spatially segregated. The thick black arrows represent ontogenetic transitions. In this context, "feedback to next generation" implies a density-dependent feedback mechanism whose target is at point of recruitment into the first juvenile stage (eggs or neonates). In such a system (in contrast to one with spatial segregation of life-stages), juvenile density effects on adult fecundity can be direct (e.g., high density of juveniles at any point in time can have an inhibitory effect on the fecundity of the adults in the population at that same time). Such effects can, therefore, give rise to a stabilizing feedback loop. Note also that adult density-dependent juvenile mortality in such sytems can produce either destabilizing or stabilizing feedback loops, depending on whether it is the older (black feedback loop in A) or younger (grey feedback loop in A) juveniles that bear the brunt of the mortality.

271

line acceleration (at low speeds) is not too high, and if the accelerator is not too sensitive to speed.

Overall, when trying to assess the impact of various aspects of a species' life history and ecology on population dynamics, one can use this heuristic framework to evaluate the potential effects of different biological factors and processes on stability and dynamics. One can examine how they map onto the ontogeny, and whether they are likely to directly affect recruitment into the first juvenile stage. Some of the major questions that need to be addressed are as follows:

- Are generations discrete or overlapping? If generations overlap, are cohorts segregated in space?
- What kinds of interactions exist among life stages? Which life stages are likely to be the triggers of density-dependent regulatory mechanisms? Often, the trigger stage is the primary consumer of resources.
- Which life stages are the likely targets of density-dependent regulatory mechanisms? If the target is the first juvenile stage, does the regulatory mechanism act primarily through fecundity or mortality?
- How do the trigger and target map onto the ontogeny, especially in the context of whether cohorts are spatially segregated? What are the time delays between triggering of a regulatory mechanism and its effect on the target, and between the effect on the target and its final effect on the triggering life stage?
- If fecundity or mortality are density independent, what is the magnitude of each?
- What is the census life stage? If this stage is not the trigger life stage, how does it map onto the ontogeny, relative to the trigger life stage, and the first juvenile stage to which recruitment is governed through fecundity?

Another point that must be taken into account with this analysis is whether the system has discrete or overlapping generations, since the stabilizing or destabilizing nature of a feedback loop is a relative notion. In populations with over-lapping generations, a long ontogenetic delay between the trigger and regulatory stages may be destabilizing, while the same delay may have the opposite effect in a discrete time system.

The important issues just listed should not be considered in isolation; real populations often have multiple census, trigger, and target life stages, as we shall see in the next section. Nevertheless, we feel that the conceptual framework outlined in this section is of value because it allows us to evaluate systematically the various aspects of the life history and ecology of a species in a manner that will be helpful in deciding which factors are most likely to affect the stability of the system.

LUCILIA CUPRINA, TRIBOLIUM AND DROSOPHILA COMPARED

In this section, we take another look at what is known about population dynamics and stability in the two best-studied model systems, *Tribolium* and *Drosophila*. We compare these species in light of the framework just discussed, highlighting similarities and differences in the way in which population growth is regulated in the two systems. We also briefly discuss a third system we examined earlier, *L. cuprina*, as an interesting contrast to *Tribolium* and *Drosophila*.

Concentrating on the latter two systems for the time being, it is clear that *Tribolium* cultures are relatively more stable with regard to adult numbers as compared with *Drosophila* cultures of about the same size subjected to a stabilizing food regime. For example, in the four control populations of *Tribolium* used by Costantino and Desharnais (1981), the mean number of adults was about 100, and the coefficient of variation of adult numbers over time was 0.21. By contrast, the mean number of adults in small *Drosophila* populations subjected to the stabilizing high adult, low larval (HL) food regime was also about 100 but the coefficient of variation of adult numbers was 0.62. There are other differences between these two systems. Figure 8.3 represents the ontogeny and life history of typical laboratory populations of *Tribolium* and *Drosophila*. By "typical," we mean the standard overlapping-generation populations of *Tribolium*, with food renewed every two weeks, used in the studies by R. F. Costantino, R. Desharnais, and coworkers (see chapter 5); and the HL type of discrete-generation *Drosophila* cultures used by L. D. Mueller, A. Joshi, and coworkers (see chapter 6). Consequently, the various types of feedback loop depicted are for parameter values seen in those typical populations. Deviation from those parameter values can lead to dynamic consequences not depicted in figure 8.3. For example, experimentally increasing the rate of adult mortality in *Tribolium* cultures can have a strongly destabilizing effect. Similarly, in a low larval and high adult food (LH) type of *Drosophila* culture, the weakening of the adult density dependent effects on female fecundity can completely eliminate the stabilizing effect of the feedback loop due to density-dependent fecundity.

Tribolium cultures have overlapping generations, and life stages or cohorts are typically not spatially segregated. In many of the *Drosophila* studies, on the other hand, populations have been maintained on fully discrete generations. Baseline fecundity in *Drosophila* is relatively high and is subject to adult density–dependent regulation. Compared with *Drosophila*, both baseline fecundity and the strength of its density-dependence are substantially lower in *Tribolium*. Moreover, the principal density-dependent regulatory mechanisms in *Tribolium* are cannibalism of eggs and pupae by adults and, to a lesser degree, cannibalism of eggs by larvae. Thus, the major trigger life stage for regulation in *Tribolium* is the adults, and the main targets of adult density–dependent mortality are eggs and pupae. At the same time,

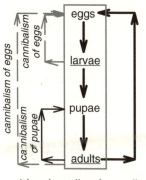
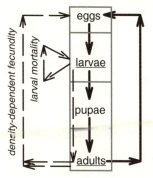

* low baseline fecundity   * high baseline fecundity

Tribolium

(overlapping generations)

Drosophila

(discrete generations)

FIG. 8.3. Schematic depiction of the ontogeny and life history of typical laboratory populations of *Tribolium* (as used in the studies by R. F. Costantino, R. Desharnais, and coworkers) and *Drosophila* (maintained on discrete generations on an HL food regime). Shown are the major density-dependent mechanisms affecting the dynamics of adult numbers. Thick black arrows represent ontogenetic transitions. Thin black solid and dashed lines indicate very strongly and moderately strongly stabilizing density-dependent feedback loops, respectively. Thin gray dashed lines indicate potentially destabilizing but weak feedback loops. Life stages that act as triggers of density-dependent regulatory mechanisms are underlined.

the larval stage is also a subsidiary trigger for cannibalism of eggs by larvae. In *Drosophila*, too, there are multiple life stages that trigger different major regulatory mechanisms. The major consumers of food are the larvae, and they act as the trigger life stage for density-dependent larval mortality. However, regulation also acts through female fecundity, and here the primary trigger life stage is the adults, although larval density also affects female fecundity because the size of an adult is reduced if it experienced relatively high densities as a larva (see fig. 8.3).

If we focus primarily on the dynamics of adult numbers, it is clear that the strongest stabilizing regulatory mechanism in *Tribolium* is adult density–dependent cannibalism of

pupae. Here, the delay between the triggering of the regulation and the impact of the pupal mortality on adult numbers
is minimal. Given the fairly high rate of pupal mortality due
to adult density–dependent cannibalism, this is a strong regulatory mechanism. Larvae in *Tribolium* do cannibalize eggs,
but the effect of this feedback on the dynamics of adult numbers, although destabilizing, is weak because it is completely
overshadowed by the feedback via adult cannibalism of
pupae. Similarly, there is also a potentially destabilizing feedback loop through adult density–dependent recruitment
into the youngest juvenile stage. Although the adult density
dependence of fecundity in *Tribolium* is very weak compared
with that of *Drosophila*, the density-dependent cannibalism
of eggs by adults plays the same role in the life history that
density-dependent fecundity does. In fact, if we examine the
strength of adult density–dependence of recruitment into
the larval stage in *Tribolium*, ignoring whether it is through
fecundity alone or a combination of fecundity and cannibalism, the recruitment falls off with increasing adult density
to a degree similar to that seen in *Drosophila* (fig. 8.4). In
an overlapping-generation culture of *Tribolium*, such adult
density dependence of larval recruitment can be a destabilizing feedback loop, but its overall impact on the dynamics of adult numbers is low because of the very low baseline
level of fecundity in *Tribolium*. As in the case of cannibalism of eggs by larvae, this effect too is overshadowed by the
strong feedback loop of adult density–dependent cannibalism of pupae.

In the case of a discrete-generation culture of *Drosophila*,
the situation is different. Here, the only main feedback loop
involving mortality is due to larval density–dependent larval mortality, and this is a stabilizing regulatory mechanism.
However, considering the dynamics of adult numbers, this
is not a strongly stabilizing loop. There is another loop that
lies between the numbers of adults in one generation and
the operation of larval density–dependent larval mortality

CONCLUSIONS

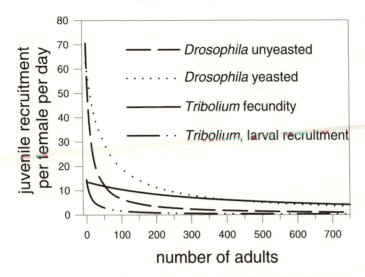

Fig. 8.4. Comparison of the sensitivity of recruitment into the first juvenile stage to adult density in laboratory populations of *Drosophila* and *Tribolium*. The figure shows best-fit curves obtained by fitting the hyperbolic model of fecundity as a function of density ($F(N_t) = a/(1+bN_t)$) to data. Data on the fecundity of *Drosophila* females at different densities (after being maintained on either yeasted or unyeasted food) were from Mueller et al. (2000) and Mueller and Huynh (1994), respectively. Data for *Tribolium* were from Rich (1956) for fecundity, and from Desharnais and Costantino (1980) for larval recruitment.

affecting adult numbers in the next generation. This intervening regulatory loop takes place through adult density–dependent female fecundity and, in a discrete-generation population that already has an inbuilt time delay, this feedback loop is stabilizing. Baseline fecundity in *Drosophila*, however, is rather high (see fig. 8.4), so this feedback loop is not as strongly stabilizing as the cannibalism of pupae in *Tribolium*. The effect of regulation through density-dependent fecundity in *Drosophila* is also strengthened to a small degree by larval density–dependent control of female fecundity. High larval density results in small adults whose baseline fecundity is, consequently, reduced. However, this effect is offset in part by the fact that high larval density

277

usually reduces subsequent adult numbers, due to high larval mortality, and this effect tends to reduce the inhibiting effect of high larval density on fecundity.

The laboratory populations of the blowfly, *Lucilia cuprina*, discussed in chapter 4, provide an interesting contrast to the *Tribolium* and *Drosophila* systems. Although the populations of *L. cuprina* used in Nicholson's (1954a, 1954b; 1957) experiments were also maintained with overlapping generations, there are several differences between them and the *Tribolium* populations we discussed in chapter 5 (Costantino et al., 1995, 1997; Dennis et al., 1995; Benoît et al., 1998). Daily adult mortality and, therefore, the rate of turnover of cohorts constituting the adult population, especially at high adult density, were far greater in *L. cuprina* than in *Tribolium*. Baseline female fecundity in *L. cuprina* was almost twice as high as in *Tribolium*, and the sensitivity to adult density of recruitment into the youngest juvenile stage was lower in *L. cuprina*, especially in the LH treatments for which adult protein supply was unlimited. Most important, both major density-dependent regulatory mechanisms in *L. cuprina*—namely, adult density–dependent female fecundity and larval density–dependent larval mortality—involved a time delay more or less equal to the development time before the triggering of the density-dependent feedback and its effect finally being felt at the adult stage. Recall here that in Nicholson's LH food regime, adults lay large numbers of eggs regardless of adult density, giving rise to high densities of newly hatched larvae. Consequently, although the regulatory mechanism of larval mortality here is, strictly speaking, triggered by larval density, the larval density itself is directly proportional to adult density a few days before. Thus, the trigger life stage here is also the adult stage. In *Tribolium*, the delay between the triggering of adult density–dependent cannibalism of pupae and its impact on adult densities is almost negligible. It is, therefore, not surprising that the dynamics of adult numbers in *Tribolium* are relatively stable, whereas in *L. cuprina* the dynamics are highly destabilized, regardless of food regime.

If we compare the effects of LH and HL food regimes on *L. cuprina* and *Drosophila*, it is clear that the HL food regime does not have the same stabilizing effect on *L. cuprina* as it does on *Drosophila*. The reason for this, we believe, is the difference in maintenance regime in the two systems. In the discrete-generation *Drosophila* populations that were subjected to HL and LH food regimes, only eggs laid during a 24-hour period were used to initiate the next generation. The *L. cuprina* populations, on the other hand, were maintained with overlapping generations, and egg laying by adults was continuous, as was recruitment into the adult life stage. When generations are fully discrete, as in the *Drosophila* populations, the HL food regime is stabilizing. This outcome occurs because the regulatory effect of adult density–dependent fecundity eventually trickles up the ontogeny, through the egg, larval, and pupal life stages, back to the adult stage that triggered it. The ontogenetic delay here is not destabilizing because, in such a system, any adult density–dependent feedback effect has to undergo this process of trickling back up through the entire ontogeny. This recruitment into the adult stage is episodic, and the adult stage cannot feed back onto preceding life stages against the flow of the ontogeny, so to speak. In a system with overlapping generations, in which recruitment to the adult stage is continuous and adult density can feed back on preadult stages against the flow of the ontogeny, a time delay in density-dependent regulatory mechanisms is destabilizing. As we noted earlier, stability is, at least in this context, a relative notion. Indeed, it is quite likely that in an overlapping generation culture of *Drosophila*, strong adult density–dependent fecundity would not in itself be a strong stabilizing factor.

Overall, we feel that the contrasts between the three model systems discussed here give support to considerations of whether generations are discrete or overlapping, and what kinds of time delays exist between the triggering and effect of density-dependent feedback mechanisms. These effects can be of great importance in determining whether a particular density-dependent mechanism will have a stabilizing or destabilizing effect on population dynamics. Some of the comparisons also exemplify the point that the same density-dependent regulatory mechanism can be either destabilizing or stabilizing, depending on the various life-history and ecological attributes of the system.

MODEL SYSTEMS IN ECOLOGY: WHERE NEXT?

Our intention in writing this book has been to review the work done on single-species population dynamics using model laboratory systems, and to highlight, through this review, the tremendous potential of such systems for testing and refining theory in population ecology. Although the 1980s and 1990s saw tremendous growth in population ecology, much of this was due to increased theoretical studies, studies on wild populations, and advances in data analysis techniques. Laboratory studies in population ecology have not registered the same kind of growth. We hope that this book helps redress this imbalance by inspiring more people to take up studies of model laboratory systems. In this last section, we highlight areas of population ecology in which studies on laboratory systems might prove especially fruitful.

As we discussed in chapter 2, there are several theoretical predictions about the impact of age structure on the dynamics of populations (Guckenheimer et al., 1977; Swick, 1981; Charlesworth, 1994). Age classes in mammals and birds are relatively easily distinguished, but these systems are not otherwise amenable to controlled experimentation. In chapter 6, we discussed one preliminary study of the effect of age structure on the dynamics of *Drosophila* populations kept under a destabilizing maintenance regime. With model systems, it should prove possible to examine critically many of the conflicting predictions from the theory.

Age-class dynamics are important to studies in both ecology and evolution. Members of different age classes may make different demands for natural resources or may show different propensities to disperse. If the absolute numbers of individuals in different age classes varies, or if their relative proportions vary, this can affect important ecological features of a population. Predictions of the effects of natural selection in populations with age structure depend on the relative distribution of individuals into different age classes (Charlesworth, 1994). The effects of natural selection on age-structured populations provide important information for understanding the evolution of iteroparity or the near-universal phenomenon of aging (Rose, 1991). It will be important in future studies to understand environmental or other factors that might cause age classes to achieve a stable age distribution or to fluctuate in a regular fashion. Such information could greatly expand the sophistication and relevance of evolutionary models with age-structured populations.

Another aspect of single-population dynamics that might benefit from studies on model laboratory systems is our understanding of the biological causes of demographic stochasticity. Stochasticity can be incorporated into models of population dynamics in many ways. One can add a stochastic component to individual parameters ot the model, or to the predicted population size in the subsequent generation. Similarly, the stochastic component may be assumed to be additive on either numerical or logarithmic scales. The impact on dynamics of incorporating stochasticity into a model in different ways has not been studied systematically. From a biological point of view, too, demographic stochasticity can arise from many causes, such as random variation in fecundity and in mortality at different life stages, or random variation in sex ratio. In chapter 6, we discussed a study with *Drosophila* suggesting that random sex-ratio variation may not be a major contributor to demographic stochasticity. It may be worthwhile to examine theoretically whether different biological causes of stochasticity translate into differences in how the stochastic component should be incorporated into a model of dynamics; and whether such differences ultimately yield varying predicted population size distributions. Empirical verification of such theory will be much easier with laboratory rather than natural systems.

The ability to manipulate migration rates and the nature of population dynamics in model systems such as *Drosophila* and *Tribolium* also makes them useful for empirical verification of various predictions from single-species metapopulation theory. Chapter 6 renewed the use of model systems to study the effects of the interplay between migration and local dynamics on overall metapopulation dynamics. There is also considerable theory about the effects of constant migration rates on dynamics, the evolution of migration rates, and the effect on local and global dynamics of migration among subpopulations that show a variety of dynamic behaviors. These are all issues that are more amenable to laboratory studies than to studies using of natural populations. Dynamics of *Tribolium* and *Drosophila* cultures can be easily manipulated, as can migration rates. In *Tribolium*, increased emigration has been successfully selected for (Goodnight, 1990a, 1990b), indicating that it may be possible to study experimentally the density-dependent evolution of migration rates.

Systems of interacting species provide examples of some of the most complex and interesting spatial and temporal patterns in dynamic behavior. Although such systems are beyond the purview of this book, we think that the use of laboratory systems would enhance our understanding of such systems. Some of the earliest empirical work on two species dynamics was done on laboratory cultures of protozoans (Gause, 1934), and much detailed empirical work on interspecific competition used laboratory populations of *Drosophila* species (Moore, 1952a, 1952b; Miller, 1964a, 1964b; Ayala, 1966, 1971; Arthur, 1980, 1986). More recently, laboratory studies have demonstrated (1) the stabilization of competitive interactions by predators (Worthen, 1989); (2) the presence of higher-order interactions and indirect effects in multispecies assemblages of competitors (Worthen and Moore, 1991); (3) a geographic mosaic in the outcomes of interspecific competition (Joshi and Thompson, 1995); and (4) the coevolution of competitors (Goodnight, 1991; Joshi and Thompson, 1996). Although this is a partial list of important experimental work in multispecies population ecology, it suffices to make the point that laboratory systems can continue to make an important contribution to our understanding of the dynamics of multispecies systems.

Some of the most interesting questions in community ecology arise in the context of multispecies metapopulations. Many of these questions, such as the effect of migration corridors on species abundance and diversity in communities, are also important to conservation biology. They have been studied using multitrophic laboratory communities of bacteria and protozoa (Burkey, 1997) and replicate field microecosystems of moss patches with their microarthropod fauna (Gonzalez et al., 1998). Fragmented microbial ecosystems with corridors went extinct significantly faster than those without corridors (Burkey, 1997), whereas in the moss patches, corridors arrested the decline in abundance and diversity of the microarthropod fauna caused by fragmentation (Gonzalez et al., 1998). Such differences in the results from different model systems highlight the need for studies on a range of model systems, which has not been the case thus far. Model microecosystems have also been used to examine the effects of productivity and patch size on food web complexity: Larger patches of habitat were found to support food webs with more species and longer food chains than were smaller patches of otherwise identical habitat (Spencer and Warren, 1996). Once again, we are citing just a few studies of this type to make the point that model systems can be useful in testing predictions from theory in community ecology as well as population ecology.

One common advantage of model systems, as opposed to the majority of field systems, is the preexisting detailed knowledge of their biology. Thus, not every population of a species that has been maintained in controlled laboratory conditions is really a model system. For well-developed model systems, a wealth of information is available on their laboratory ecology, life history, and genetics, and it is this knowledge that allows an experimenter to go beyond observing certain dynamic behaviors and analyzing the observed patterns in time and space. Thus, as we have seen in the case of *Drosophila* and *Tribolium*, we now actually understand a great deal about how particular ecological or life-history phenomena give rise to certain kinds of dynamic behavior. One of the drawbacks with most natural systems is that, even if our knowledge of their ecology is reasonably good, we typically know very little about the genetic architecture of fitness components in those environments. Yet, as population ecology and population genetics move closer to one another, many of the interesting questions in evolution and ecology lie on their interface. Model systems such as *Drosophila* are extremely useful for addressing these questions because their laboratory ecology and evolutionary genetics have been studied extensively under a variety of laboratory environments. Questions about the evolution of population dynamics and stability, involving an evaluation of hypotheses based on group selection versus individual selection, are amenable to empirical study with model systems. So are issues such as the impact of inbreeding levels on population extinction rates, the genetic effective size of metapopulations, and the determination of minimum viable metapopulations from demographic and genetic points of view.

Ultimately, our understanding of population or community dynamics will be best served by a three-pronged approach that involves feedback from theory, laboratory experiments, and field studies. For this approach to be balanced, we need more studies on model systems as well as a greater diversity of model systems that are well characterized both ecologically and genetically. At this time, *Drosophila* and *Tribolium* remain the best model systems for studies in population ecology. Genetically, *Drosophila* is better characterized, but rapid advances have been made in understanding *Tribolium* genetics (e.g., Alvarez-Fuster et al., 1991; Beeman et al., 1996; Beeman and Brown, 1999). Other laboratory systems have already been used extensively in studies on evolutionary genetics and life-history evolution; These include bacteria (Lenski and Travisano, 1994; Vasi et al., 1994; Travisano et al., 1995a, 1995b; Elena et al., 1996; Elena and Lenski, 1997), the bruchids *Callosobruchus* (Møller et al., 1989; Tatar et al., 1993; Tatar and Carey, 1994, 1995), and *Acanthoscelides* (Tucic et al., 1990, 1996, 1997). Some of these systems might be used fruitfully in future investigations of population dynamics.

References

Adler, F. R. 1993. Migration alone can produce persistence of host-parasitoid models. *American Naturalist* 141: 642–650.

Aiken, R. B., and D. L. Gibo. 1979. Changes in fecundity of *Drosophila melanogaster* and *D. simulans* in response to selection for competitive ability. *Oecologia* 43: 63–77.

Allen, D. M. 1974. The relationship between variable selection and data augmentation and a method for prediction. *Technometrics* 16: 125–127.

Allen, J. C., W. M. Schaffer, and D. Rosko. 1993. Chaos reduces species extinction by amplifying local noise. *Nature* 364: 229–232.

Allendorf, F. W., and R. F. Leary. 1986. Heterozygosity and fitness in natural populations of animals. In *Conservation Biology: The Science of Scarcity and Diversity*, ed. M. E. Soulé, 57–76. Sinauer, Sunderland, Massachusetts.

Alvarez-Fuster, A., C. Juan, and E. Petitpierre. 1991. Genome size in *Tribolium* flour beetles: Inter- and intraspecific variation. *Genetical Research* 58: 1–5.

Amarasekare, P. 1998. Interactions between local dynamics and dispersal: Insights from single species models. *Theoretical Population Biology* 53: 44–59.

Anderson W. W. 1971. Genetic equilibrium and population growth under density-regulation selection. *American Naturalist* 105: 489–498.

Andrewartha, H. G., and L. C. Birch. 1954. *The Distribution and Abundance of Animals*. Univ. of Chicago Press, Illinois.

Ariño, A., and S. L. Pimm. 1995. On the nature of population extremes. *Evolutionary Ecology* 9: 429–443.

Arthur, W. 1980. Interspecific competition in *Drosophila*. I. Reversal of competitive superiority due to varying concentration of ethanol. *Biological Journal of the Linnean Society* 13: 109–118.

Arthur, W. 1986. On the complexity of a single environment: Competition, resource partitioning and facilitation in a two-species *Drosophila* system. *Philosophical Transactions of the Royal Society of London B* 313: 471–508.

Asmussen, M. A. 1983. Density-dependent selection incorporating intraspecific competition. II. A diploid model. *Genetics* 103: 335–350.

287

REFERENCES

Ayala, F. J. 1965a. Relative fitness of populations of *Drosophila serrata* and *Drosophila birchii*. *Genetics* 51: 527–544.

Ayala, F. J. 1965b. Evolution of fitness in experimental populations of *Drosophila serrata*. *Science* 150: 903–905.

Ayala, F. J. 1966. Reversals of dominance in competing species of *Drosophila*. *American Naturalist* 100: 81–83.

Ayala, F. J. 1968. Genotype, environment, and population numbers. *Science* 162: 1453–1459.

Ayala, F. J. 1969. Experimental invalidation of the principle of competitive exclusion. *Nature* 224: 1076–1079.

Ayala, F. J. 1971. Competition between strains of *Drosophila willistoni* and *D. pseudoobscura*. *Experientia* 27: 343.

Ayala, F. J., M. E. Gilpin, and J. G. Ehrenfeld. 1973. Competition between species: Theoretical models and experimental tests. *Theoretical Population Biology* 4: 331–356.

Bakker, K. 1961. An analysis of factors which determine success in competition for food among larvae of *Drosophila melanogaster*. *Archives neerlandaises de zoologie*. 14: 200–281.

Barker, J.S.F. 1974. Ecological differences and competitive interaction between *Drosophila melanogaster* and *Drosophila simulans* in small laboratory populations. *Oecologia* 8: 139–156.

Barlow, J. 1992. Nonlinear and logistic growth in experimental populations of guppies. *Ecology* 73: 941–950.

Batzli, G. O. 1992. Dynamics of small mammal populations: A review. In *Wildlife 2001: populations*, ed. D. R. McCullogh and R. II. Barrett, 831–850. Elsevier Applied Sciences, London.

Beeman, R. W., and S. J. Brown. 1999. RAPD-based genetic linkage maps of *Tribolium castaneum*. *Genetics* 153: 333–338.

Beeman, R. W., J. J. Stuart, M. S. Haas, and K. S. Friesen. 1996. Chromosome extraction and revision of linkage group 2 in *Tribolium castaneum*. *Journal of Heredity* 87: 224–232.

Begon M., J. L. Harper, and C. R. Townsend. 1990. *Ecology, Individuals, Populations and Communities*. Blackwell, Cambridge, UK.

Benoît, H. P., E. McCauley, and J. R. Post. 1998. Testing the demographic consequences of cannibalism in *Tribolium confusum*. *Ecology* 79: 2839–2851.

Berryman, A. A., and J. A. Millstein. 1989. Are ecological systems chaotic—and if not, why not? *Trends in Ecology and Evolution* 4: 26–28.

Beveridge, G.S.G., and R. S. Schechter. 1970. *Optimization: Theory and Practice*. McGraw-Hill, New York.

288

REFERENCES

Bijlsma, R., J. Bundgaard, and A. C. Boerema. 2000. Does inbreeding affect the extinction risk of small populations?: Predictions from *Drosophila*. *Journal of Evolutionary Biology* 13: 502–514.

Bijlsma, R., J. Bundgaard, A. C. Boerema, and W. F. Van Putten. 1997. Genetic and environmental stress, and the persistence of populations. In *Environmental Stress, Adaptation and Evolution*, ed. R. Bijlsma and V. Loeschcke, 193–207. Birkhauser Verlag, Basel, Switzerland.

Borash, D. J., A. G. Gibbs, A. Joshi, and L. D. Mueller. 1998. A genetic polymorphism maintained by natural selection in a temporally varying environment. *American Naturalist* 151: 148–156.

Borash, D. J., H. Teotónio, M. R. Rose, and L. D. Mueller. 2000. Density-dependent natural selection in *Drosophila*: Correlations between feeding rate, development time, and viability. *Journal of Evolutionary Biology* 13: 181–187.

Brillinger, D. R., J. Guckenheimer, P. Guttorp, and G. Oster. 1980. Empirical modelling of population time series data: The case of age and density dependent vital rates. *Lectures on Mathematics in the Life Sciences* 13: 65–90.

Bundgaard, J., and F. B. Christiansen. 1972. Dynamics of polymorphisms. I. Selection components in an experimental population of *Drosophila melanogaster*. *Genetics* 71: 439–460.

Burkey, T. V. 1997. Metapopulation extinction in fragmented landscapes: Using bacteria and protozoa communities as model ecosystems. *American Naturalist* 150: 568–591.

Burnet B., D. Sewell, and M. Bos. 1977. Genetic analysis of larval feeding behavior in *Drosophila melanogaster*. II. Growth relations and competition between selected lines. *Genetical Research* 30: 149–161.

Buzzati-Traverso, A. A. 1955. Evolutionary changes in components of fitness and other polygenic traits in *Drosophila melanogaster* populations. *Heredity* 9: 153–186.

Cappucino, N., and P. W. Price (eds). 1995. *Population Dynamics: New Approaches and Synthesis*. Academic Press, San Diego, California.

Caro, T. M., and M. K. Laurenson. 1994. Ecological and genetic factors in conservation: A cautionary tale. *Science* 263: 485–486.

Carpenter, S. R. 1996. Microcosm experiments have limited relevance for community and ecosystem ecology. *Ecology* 77: 677–680.

Caswell, H., and J. E. Cohen. 1991. Disturbance, interspecific interaction and diversity in metapopulations. *Biological Journal of the Linnean Society* 42: 193–218.

Cavalieri, L. F., and H. Koçak. 1995. Intermittent transition between order and chaos in an insect pest population. *Journal of Theoretical Biology* 175: 231–234.

Chapman, R. N. 1918. The confused flour beetle (*Tribolium confusum* Duval). *Minnesota State Entomological Report* 17: 73–94.

Chapman, R. N. 1928. Quantitative analysis of environmental factors. *Ecology* 9: 111–122.

Chapman, R. N. 1933. The causes of fluctuations of populations of insects. *Proceedings of the Hawaiian Entomological Society* 8: 279–297.

Chapman, R. N., and L. Baird. 1934. The biotic constants of *Tribolium confusum* Duval. *Journal of Experimental Zoology* 68: 293–305.

Chapman, R. N., and W. Y. Whang. 1934. An experimental analysis of the cause of population fluctuations. *Science* 80: 297–298.

Charlesworth, B. 1971. Selection in density-regulated populations. *Ecology* 52: 469–474.

Charlesworth, B. 1994. *Evolution in Age-Structured Populations.* 2nd ed. Cambridge Univ. Press, London.

Chatfield, C. 1989. *The analysis of time series.* 4th ed. Chapman & Hall, London.

Chiang, H. C., and A. G. Hodson. 1950. An analytical study of population growth in *Drosophila melanogaster. Ecological Monographs* 20: 173–206.

Christiansen, F. B. 1984. Evolution in temporally varying environments: Density and composition dependent genotypic fitness. In *Population Biology and Evolution,* ed. K. Wöhrmann and V. Loeschcke, 115–124. Springer-Verlag, Berlin.

Clarke, B. 1972. Density-dependent selection. *American Naturalist* 106: 1–13.

Clutton-Brock, T. H., A. W. Illius, K. Wilson, B. T. Grenfell, A.D.C. MacColl, and S. D. Albon. 1997. Stability and instability in ungulate populations: An empirical analysis. *American Naturalist* 149: 195–219.

Cohen, J. E. 1995. Unexpected dominance of high frequencies in chaotic nonlinear population models. *Nature* 378: 610–612.

Comins, H. N., and M. P. Hassell. 1996. Persistence of multispecies host-parasitoid interactions in spatially distributed models with local dispersal. *Journal of Theoretical Biology* 183: 19–28.

Connell, J. H., and W. P. Sousa. 1983. On the evidence needed to judge ecological stability or persistence. *American Naturalist* 121: 789–824.

REFERENCES

Costantino, R. F., and R. A. Desharnais. 1981. Gamma distributions of adult numbers for *Tribolium* populations in the region of their steady states. *Journal of Animal Ecology* 50: 667–681.

Costantino, R. F., and R. A. Desharnais. 1991. *Population Dynamics and the* Tribolium *Model: Genetics and Demography.* Springer-Verlag, New York.

Costantino, R. F., J. M. Cushing, B. Dennis, and R. A. Desharnais. 1995. Experimentally induced transitions in the dynamic behaviour of insect populations. *Nature* 375: 227–230.

Costantino, R. F., R. A. Desharnais, J. M. Cushing, and B. Dennis. 1997. Chaotic dynamics in an insect population. *Science* 275: 389–391.

Crombie, A. C. 1946. Further experiments on insect competition. *Proceedings of the Royal Society of London B* 133: 76–109.

Curtsinger, J. W., H. H. Fukui, A. A. Khazaeli, A. Kirscher, S. D. Pletcher, D.E.L. Promislow, and M. Tatar. 1995. Genetic variation and aging. *Annual Review of Genetics* 29: 553–575.

Cushing, J. M. 1977. *Integrodifferential Equations and Delay Models in Population Dynamics.* Springer-Verlag, Berlin.

Darwin, C. 1859. *The Origin of Species.* Penguin, London.

Dawson, P. S. 1964. An interesting behavioral phenomenon in *Tribolium confusum. Tribolium Information Bulletin* 7: 50–52.

Dawson, P. S. 1965. Genetic homoeostasis and developmental rate in *Tribolium. Genetics* 51: 873–885.

De Belle, J. S., A. J. Hilliker, and M. B. Sokolowski. 1989. Genetic localization of *foraging (for)*: A major gene for larval behavior in *Drosophila melanogaster. Genetics* 123: 157–163.

Del Solar, E., and H. Palomino. 1966. Choice of oviposition in *Drosophila melanogaster. American Naturalist* 100: 127–133.

Dempster, J. P. 1983. The natural control of populations of butterflies and moths. *Biological Review of the Cambridge Philosophical Society* 58: 461–481.

Den Boer, P. J. 1968. Spreading of risk and stabilization of animal numbers. *Acta Biotheoretica* 18: 165–194.

Dennis, B., and R. F. Costantino. 1988. Analysis of steady-state populations with the Gamma abundance model and its application to *Tribolium. Ecology* 69: 1200–1213.

Dennis, B., R. A. Desharnais, J. M. Cushing, and R. F. Costantino. 1995. Nonlinear demographic dynamics: Mathematical models, statistical methods, and biological experiments. *Ecological Monographs* 65: 261–281.

291

Dennis, B., R. A. Desharnais, J. M. Cushing, and R. F. Costantino. 1997. Transitions in population dynamics: Equilibria to periodic cycles to aperiodic cycles. *Journal of Animal Ecology* 66: 704–729.

Dennis, B., R. A. Desharnais, J. M. Cushing, S. M. Henson, and R. F. Costantino. (in press). Estimating chaos and complex dynamics in an insect population. *Ecology.*

Dennis, B., P. L. Munholland, and J. M. Scott. 1991. Estimation of growth and extinction parameters for endangered species. *Ecological Monographs* 61: 115–143.

Dennis, B., and M. L. Taper. 1994. Density-dependence in time series observations of natural populations: Estimation and testing. *Ecological Monographs* 64: 205–224.

Desharnais, R. A., and R. F. Costantino. 1980. Genetic analysis of a population of *Tribolium*. VII. Stability: Response to genetic and demographic perturbations. *Canadian Journal of Genetics and Cytology* 22: 577–589.

Desharnais, R. A., and R. F. Costantino. 1982. The approach to equilibrium and the steady state probability distribution of adult numbers in *Tribolium brevicornis*. *American Naturalist* 119: 102–111.

Desharnais, R. A., R. F. Costantino, J. M. Cushing, and B. Dennis. 1997. Response to technical comment by J. N. Perry et al. *Science* 276: 1881–1882.

Desharnais, R. A., and L. Liu. 1987. Stable demographic limit cycles in laboratory populations of *Tribolium castaneum*. *Journal of Animal Ecology* 56: 885–906.

Doebeli, M. 1994. Intermittent chaos in population dynamics. *Journal of Theoretical Biology* 166: 325–330.

Doebeli, M., and G. de Jong. 1999. Genetic variability in sensitivity to population density affects the dynamics of simple ecological models. *Theoretical Population Biology* 55: 37–52.

Doebeli, M., and J. C. Koella. 1995. Evolution of simple population dynamics. *Proceedings of the Royal Society of London* B 260: 119–125.

Doebeli, M., and J. C. Koella. 1996. Chaos and evolution. *Trends in Ecology and Evolution* 11: 220.

Doebeli, M., and G. D. Ruxton. 1998. Stabilization through spatial pattern formation in metapopulations with long-range dispersal. *Proceedings of the Royal Society of London* B 265: 1325–1332.

Drake, J. A., G. R. Huxel, and C. L. Hewitt. 1996. Microcosms as models for generationg and testing community theory. *Ecology* 77: 670–677.

Earn, D.J.D., P. Rohani, and B. T. Grenfell. 1998. Persistence, chaos and synchrony in ecology and epidemiology. *Proceedings of the Royal Society of London* B 265: 7–10.

REFERENCES

Efron, B., and R. J. Tibshirani. 1993. *An Introduction to the Bootstrap.* Chapman & Hall, New York.

Elena, S. F., V. S. Cooper, and R. E. Lenski. 1996. Punctuated evolution caused by selection of rare beneficial mutations. *Science* 272: 1802–1804.

Elena, S. F., and R. E. Lenski. 1997. Long-term experimental evolution in *Escherichia coli.* VII. Mechanisms maintaining genetic variability within populations. *Evolution* 51: 1058–1067.

Ellner, S., and P. Turchin. 1995. Chaos in a noisy world: New methods and evidence from time-series analysis. *American Naturalist* 145: 343–375.

Elton, C. S. 1924. Periodic fluctuations in the number of animals: Their causes and effects. *British Journal of Experimental Biology* 2: 119–163.

Englert, D. S., and A. E. Bell. 1970. Selection for time of pupation in *Tribolium castaneum. Genetics* 64: 541–552.

Ewens, W. J. 1979. *Mathematical Population Genetics.* Springer-Verlag, Berlin.

Ewens, W. J., P. J. Brockwell, J. M. Gani, and S. I. Resnick. 1987. Minimum viable population size in the presence of catastrophes. In *Viable Populations for Conservation,* ed. M. E. Soulé, 59–68. Cambridge Univ. Press, New York.

Falck, W., O. N. Bjørnstad, and N. C. Stenseth. 1995a. Bootstrap estimated uncertainty of the dominant Lyapunov exponent for Holarctic microtine rodents. *Proceedings of the Royal Society of London B* 261: 159–165.

Falck, W., O. N. Bjørnstad, and N. C. Stenseth. 1995b. Voles and lemmings: Chaos and uncertainty in fluctuating populations. *Proceedings of the Royal Society of London B* 262: 363–370.

Ferrière, R., and G. A. Fox. 1995. Chaos and evolution. *Trends in Ecology and Evolution* 10: 480–485.

Forney, K. A., and M. E. Gilpin. 1989. Spatial structure and population extinction: A study with *Drosophila* flies. *Conservation Biology* 3: 45–51.

Fox, G. A. 1996. Chaos and evolution. *Trends in Ecology and Evolution* 11: 336–337.

Futuyma, D. J. 1970. Variation in genetic response to interspecific competition in laboratory populations of *Drosophila. American Naturalist* 104: 239–252.

Gadgil, M. 1971. Dispersal: Population consequences and evolution. *Ecology* 52: 253–261.

Gallant, A. R. 1975. Nonlinear regression. *American Statistician* 29: 73–81.

Gatto, M. 1993. The evolutionary optimality of oscillatory and chaotic dynamics in simple population models. *Theoretical Population Biology* 43: 310–336.

Gause, G. F. 1934. *The Struggle for Existence.* Dover, New York.

Gillespie, J. H. 1974. Natural selection for within-generation variance in offspring number. *Genetics* 76: 601–606.

Gilpin, M. E. 1975. *Group Selection in Predator-Prey Communities.* Princeton Univ. Press, Princeton, New Jersey.

Gilpin, M. E. 1992. Demographic stochasticity: A Markovian approach. *Journal of Theoretical Biology* 154: 1–8.

Gilpin, M. E., and F. J. Ayala. 1973. Global models of growth and competition. *Proceedings of the National Academy of Sciences USA* 70: 3590–3593.

Gilpin, M. E., and I. Hanski. 1991. *Metapopulation Dynamics: Empirical and Theoretical Investigations.* Academic Press, San Diego, California.

Gilpin, M. E., and I. Hanski. 1997. *Metapopulation Biology: Ecology, Genetics and Evolution.* Academic Press, San Diego, California.

Goldberg, S. 1958. *Difference Equations.* John Wiley & Sons, New York.

Gomulkiewicz, R., R. D. Holt, and M. Barfield. 1999. The effects of density dependence and immigration on local adaptation and niche adaptation and niche evolution in a black-hole sink environment. *Theoretical Population Biology* 55: 283–296.

Gonzalez, A., J. H. Lawton, F. S. Gilbert, T. M. Blackburn, and I. Evans-Freke. 1998. Metapopulation dynamics, abundance and distribution in a microecosystem. *Science* 281: 2045–2047.

Goodnight, C. J. 1990a. Experimental studies of community evolution. I. The response to selection at the community level. *Evolution* 44: 1614–1624.

Goodnight, C. J. 1990b. Experimental studies of community evolution. II. The ecological basis of the response to community selection. *Evolution* 44: 1625–1636.

Goodnight, C. J. 1991. Intermixing ability in two-species communities of *Tribolium* flour beetles. *American Naturalist* 138: 342–354.

Grenfell, B. T., K. Wilson, B. M. Finkenstädt, T. N. Coulson, S. Murray, S. D. Albon, J. M. Pemberton, T. H. Clutton-Brock, and M. J. Crawley. 1998. Noise and determinism in synchronized sheep dynamics. *Nature* 394: 674–677.

Guckenheimer, J., G. F. Oster, and A. Ipaktchi. 1977. The dynamics of density-dependent population models. *Journal of Mathematical Biology* 4: 101–147.

REFERENCES

Guo, P. Z., L. D. Mueller, and F. J. Ayala. 1991. Evolution of behavior by density-dependent natural selection. *Proceedings of the National Academy of Sciences USA* 88: 10905–10906.

Gurney, W.S.C., S. P. Blythe, and R. M. Nisbet.1980. Nicholson's blowflies revisited. *Nature* 287: 17–21.

Gurney, W.S.C., and R. M. Nisbet. 1985. Fluctuation periodicity, generation separation, and the expression of larval competition. *Theoretical Population Biology* 28: 150–180.

Gutierrez, A. P. 1996. *Applied Population Ecology: A Supply Demand Approach*. Wiley, New York.

Gyllenberg, M., G. Söderbacka, and S. Ericsson. 1993. Does migration stabilise local dynamics? Analysis of a discrete metapopulation model. *Mathematical Biosciences* 118: 25–49.

Hansen, T. F. 1992. Evolution of stability parameters in single species population models: Stability or chaos? *Theoretical Population Biology* 42: 199–217.

Hanski, I., and M. E. Gilpin. 1997. *Metapopulation Biology: Ecology, Genetics, and Evolution*. Academic Press, San Diego, California.

Hanski, I., and M. Gyllenberg. 1993. Two general metapopulation models and the core-satellite species hypothesis. *American Naturalist* 142: 17–41.

Hanski, I., M. Kuussaari, and M. Nieminen. 1994. Metapopulation structure and migration in the butterfly *Melitaea cinxia*. *Ecology* 75: 747–762.

Hanski, I., and D. Simberloff. 1997. The metapopulation approach, its history, conceptual domain, and application to conservation. In *Metapopulation Biology: Ecology, Genetics, and Evolution*, ed. I. Hanski and M. E. Gilpin, 27–42. Academic Press, San Diego, California.

Hanski, I., and D. Y. Zhang. 1993. Migration, metapopulation dynamics, and fugitive co-existence. *Journal of Theoretical Biology* 163: 491–504.

Harrison, S. 1994. Metapopulations and conservation. In *Large-Scale Ecology and Conservation Biology*, ed. P. J. Edwards, N. R. Webb, and R. M. May, 111–128. Blackwell Scientific, Oxford, UK.

Harrison, S., and A. D. Taylor. 1997. Empirical evidence for metapopulation dynamics. In *Metapopulation Biology: Ecology, Genetics, and Evolution*, ed. I. Hanski and M. E. Gilpin, 5–26. Academic Press, San Diego, California.

Hassell M., J. Lawton, and R. M. May. 1976. Pattern of dynamical behavior in single species populations. *Journal of Animal Ecology* 45: 471–486.

Hassell, M. P., O. Miramontes, P. Rohani, and R. M. May. 1995. Appropriate formulations for dispersal in spatially structured dynamics: Comments on Bascompte & Solé. *Journal of Animal Ecology* 64: 662–664.

Hastings, A. 1991. Structured models of metapopulation dynamics. *Biological Journal of the Linnean Society* 42: 57–71.

Hastings, A. 1993. Complex interactions between dispersal and dynamics: Lessons from coupled logistic equations. *Ecology* 74: 1362–1372.

Hastings, A. 1997. *Population Biology: Concepts and Models.* Springer, New York.

Hastings, A., and R. F. Costantino. 1987. Oscillations in population numbers: Age-dependent cannibalism. *Journal of Animal Ecology* 60: 3471–3482.

Hastings, A., and R. F. Costantino. 1991. Cannibalistic egg-larva interactions in Tribolium: An explanation for the oscillations in population numbers. *American Naturalist* 130: 36–52.

Hastings, A., and K. Higgins. 1994. Persistence of transients in spatially structured ecological models. *Science* 263: 1133–1136.

Hastings, A., J. M. Serradilla, and F. J. Ayala. 1981. Boundary-layer model for the population dynamics of single species. *Proceedings of the National Academy of Sciences USA* 78: 1972–1975.

Hastings, A., and C. L. Wolin. 1989. Within-patch dynamics in a metapopulation. *Ecology* 70: 1261–1266.

Haydon, D., and H. Steen. 1997. The effects of large- and small-scale random events on the synchrony of metapopulation dynamics: A theoretical analysis. *Proceedings of the Royal Society of London B* 264: 1375–1381.

Heckel, D. G., and J. Roughgarden. 1980. A species near equilibrium size in a fluctuating environment can evolve a lower intrinsic rate of increase. *Proceedings of the National Academy of Sciences USA* 77: 7497–7500.

Hedrick, P. W. 1972. Factors responsible for a change in interspecific competitive ability in *Drosophila*. *Evolution* 26: 513–522.

Hilborn, R. C. 1994. *Chaos and Nonlinear Dynamics.* Oxford Univ. Press, New York.

Holt, R. D., and M. A. McPeek. 1996. Chaotic population dynamics favors the evolution of dispersal. *American Naturalist* 148: 709–718.

Howard, L. O., and W. F. Fiske. 1911. The importation into the United States of the parasites of the gypsy moth and the brown-tail moth. *Bulletin of the USDA Bureau of Entomology* 91.

Howe, R. W. 1956. The effects of temperature and humidity on the rate of development and mortality of *Tribolium castaneum* (Herbst.) (Coleoptera, Tenebrionidae). *Annals of Applied Biology* 44: 356–368.

Howe, R. W. 1960. The effects of temperature and humidity on the oviposition rate of *Tribolium castaneum* (Hbst.) (Coleoptera, Tenebrionidae). *Bulletin of Entomological Research* 53: 301–310.

Howe, R. W. 1962. The effects of temperature and humidity on the rate of development and mortality of *Tribolium confusum* Duval (Coleoptera, Tenebrionidae). *Annals of Applied Biology* 48: 363–376.

Howe, R. W. 1963. The prediction of the status of a pest by means of laboratory experiments. *World Review of Pest Control* 2: 2–12.

Hudson, P. J., A. P. Dobson, and D. Newborn. 1998. Prevention of population cycles by parasite removal, *Science* 282: 2256–2258.

Hudson, P. J., A. P. Dobson, and D. Newborn. 1999. Population cycles and parasitism. *Science* 286: 2425.

Huffaker, C. B. 1958. Experimental studies of predation: Dispersal factors and predator-prey oscillation. *Hilgardia* 27: 343–383.

Ims, R. A., and N. G. Yoccoz. 1997. Studying transfer processes in metapopulations: Emigration, migration and colonization. In *Metapopulation Biology: Ecology, Genetics, and Evolution,* ed. I. Hanski and M. E. Gilpin, 247–265. Academic Press, San Diego, California.

Intriligator, M. D. 1971. *Mathematical Optimization and Economic Theory.* Prentice Hall, Englewood Cliffs, New Jersey.

Jenkins, G. M., and D. G. Watts. 1968. *Spectral Analysis and Its Applications.* Holden-Day, San Francisco.

Joshi, A., and L. D. Mueller. 1988. Evolution of higher feeding rate in *Drosophila* due to density-dependent natural selection. *Evolution* 42: 1090–1093.

Joshi, A., and L. D. Mueller. 1993. Directional and stabilizing density-dependent natural selection for pupation height in *Drosophila melanogaster. Evolution* 47: 176–184.

Joshi, A., W. A. Oshiro, J. Shiotsugu, and L. D. Mueller. 1998. Short- and long-term effects of environmental urea on fecundity in *Drosophila melanogaster. Journal of Biosciences* 23: 279–283.

Joshi, A., J. Shiotsugu, and L. D. Mueller. 1996. Phenotypic enhancement of longevity by environmental urea in *Drosophila melanogaster. Experimental Gerontology* 31: 533–544.

Joshi, A., and J. N. Thompson. 1995. Alternative routes to the evolution of competitive ability in two competing species of *Drosophila. Evolution* 49: 616–625.

REFERENCES

Joshi, A., and J. N. Thompson. 1996. Evolution of broad and specific competitive ability in novel versus familiar environments in *Drosophila* species. *Evolution* 50: 188–194.

Kareiva, P. 1987. Habitat fragmentation and the stability of predator-prey interactions. *Nature* 326: 388–390.

Kareiva, P. 1990. Population dynamics in spatially complex environments. *Philosophical Transactions of the Royal Society of London B* 330: 175–190.

Kellert, Stephen H. 1993. *In the Wake of Chaos*. Univ. of Chicago Press, Illinois.

Kendall, B. E., C. J. Briggs, W. W. Murdoch, P. Turchin, S. P. Ellner, E. McCauley, R. M. Nisbet, and S. N. Wood. 1999. Why do populations cycle? A synthesis of statistical and mechanistic modeling approaches. *Ecology* 80: 1789–1805.

Kendall, M. G., and A. Stuart. 1969. *The Advanced Theory of Statistics*. Vol. I. Hafner, New York.

King, C. E., and P. S. Dawson. 1972. Population biology and the *Tribolium* model. *Evolutionary Biology* 5: 133–227.

Kingman, J.F.C. 1961. A matrix inequality. *Quarterly Journal of Mathematics* 12: 78–80.

Laing, J. E., and C. B. Huffaker. 1969. Comparative studies of predation by *Phytoseiulus persimilis* Athias-Henriot and *Metaseiulus occidentalis* (Nesbitt) (Acarina: Phytoseiidae) on populations of *Tetranychus urticae* Koch (Acarina: Tetranychidae). *Researches on Population Ecology* 11: 105–126.

Lambin, X., C. J. Krebs, R. Moss, N. C. Stenseth, and V. G. Yoccoz. 1999. Population cycles and parasitism. *Science* 286: 2425.

Lande, R. 1988. Genetics and demography in biological conservation. *Science* 241: 1455–1460.

Lande, R. 1993. Risks of population extinction from demographic and environmental stochasticity and random catastrophes. *American Naturalist* 142: 911–927.

Lande, R., and G. F. Barrowclough. 1987. Effective population size, genetic variation, and their use in population management. In *Viable Populations for Conservation*, ed. M. E. Soulé, 87–123. Cambridge Univ. Press, New York.

Leigh, E. G. 1981. The average life-time of a population in a varying environment. *Journal of Theoretical Biology* 90: 213–239.

Lenski, R. E., and M. Travisano. 1994. Dynamics of adaptation and diversification: A 10,000 generation experiment with bacterial populations. *Proceedings of the National Academy of Sciences USA* 91: 6808–6814.

REFERENCES

Lerner, I. M., and F. K. Ho. 1961. Genotype and competitive ability of *Tribolium* species. *American Naturalist* 95: 329–343.

Leslie, P. H. 1962. A stochastic model for two competing species of *Tribolium* and its application to some experimental data. *Biometrika* 49: 1–25.

Levins, R. 1968. *Evolution in Changing Environments*. Princeton Univ. Press, Princeton, New Jersey.

Levins, R. 1969. Some demographic and genetic consequences of environmental heterogeneity for biological control. *Bulletin of the Entomological Society of America* 15: 237–240.

Levins, R. 1970. Extinction. In *Some Mathematical Problems in Biology*, ed. M. Gerstenhaber, 77–107. Mathematical Society, Providence, Rhode Island.

Levins, R., and D. Culver. 1971. Regional coexistence of species and competition between rare species. *Proceedings of the National Academy of Sciences USA* 68: 1246–1248.

Lewis, O. T., C. D. Thomas, J. K. Hill, M. I. Brookes, T.P.R. Crane, Y. A. Graneau, J. B. Mallet, and O. C. Rose. 1997. Three ways of assessing metapopulation structure in the butterfly *Plebejus argus*. *Ecological Entomology* 22: 283–293.

Lewontin, R. C. 1969. The meaning of stability. In *Diversity and Stability in Ecological Systems*, Brookhaven Symp. Biol. No. 22, Springfield, VA.

Lewontin, R. C. 1974. *The Genetic Basis of Evolutionary Change*. Columbia Univ. Press, New York.

Livdahl, T. P., and G. Sugihara. 1984. Non-linear interactions of populations and the importance of estimating per capita rates of change. *Journal of Animal Ecology* 53: 573–580.

Lomnicki, A., 1988. *Population Ecology and Individuals*. Princeton Univ. Press, Princeton, New Jersey.

Ludwig, D. 1975. Persistence of dynamics systems under random perturbations. *Society for Industrial and Applied Mathematics Review* 17: 605–640.

Ludwig, D. 1976. A singular perturbation problem in the theory of population extinction. *Society for Industrial and Applied Mathematics—American Mathematical Society Proceedings* 10: 87–104.

Ludwig, D. 1996. The distribution of population survival times. *American Naturalist* 147: 506–526.

Ludwig, D. 1999. Is it meaningful to estimate a probability of extinction? *Ecology* 80: 298–310.

MacArthur, R. H. 1962. Some generalized theorems of natural selection. *Proceedings of the National Academy of Sciences USA* 48: 1893–1897.

REFERENCES

MacArthur, R. H., and E. O. Wilson. 1967. *The Theory of Island Biogeography.* Princeton Univ. Press, Princeton, New Jersey.

MacKensie, J.M.D. 1952. Fluctuations in the numbers of British Tetraonids. *Journal of Animal Ecology* 21: 128–153.

Mallows, C. L. 1973. Some comments on C_p. *Technometrics* 15: 661–675.

Mangel, M., and C. Tier. 1993. A simple direct method for finding persistence times of populations and application to conservation problems. *Proceedings of the National Academy of Sciences USA* 90: 1083–1086.

Mangel, M., and C. Tier. 1994. Four facts every conservation biologist should know about persistence. *Ecology* 75: 607–614.

Manly, B.F.J. 1990. *Stage-Structured Populations: Sampling, Analysis and Simulation.* Chapman and Hall, London.

Marinkovic, D. 1967. Genetic loads affecting fecundity in natural populations of *D. pseudoobscura. Genetics* 56: 61–71.

Marino, P. C. 1991a. Competition between mosses (Splachnaceae) in patchy habitats. *Journal of Ecology* 79: 1031–1046.

Marino, P. C. 1991b. Dispersal and coexistence of mosses (Splachnaceae) in patchy habitats. *Journal of Ecology* 79: 1047–1060.

May, R. M. 1973. *Stability and Complexity in Model Ecosytems.* Princeton Univ. Press, Princeton, New Jersey.

May, R. M. 1974. Biological populations with non-overlapping generations: Stable points, stable cycles and chaos. *Science* 186: 645–647.

May, R. M., and G. F. Oster. 1976. Bifurcations and dynamic complexity in simple ecological models. *American Naturalist* 110: 573–599.

McCallum, H. I. 1992. Effects of immigration on chaotic population dynamics. *Journal of Theoretical Biology* 154: 277–284.

McNair, J. N. 1989. Stability effects of a juvenile period in age-structured populations. *Journal of Theoretical Biology* 137: 397–422.

McNair, J. N. 1995. Ontogenetic patterns of density-dependent mortality: Contrasting stability effects in populations with adult dominance. *Journal of Theoretical Biology* 175: 207–230.

Mertz, D. B., and R. B. Davies. 1968. Cannibalism of the pupal stage by adult flour beetles: An experiment and stochastic model. *Biometrics* 24: 247–275.

Middleton, A. D. 1934. Periodic fluctuations in British game populations. *Journal of Animal Ecology* 3: 231–249.

Miller, R. J. 1974. The jackknife—a review. *Biometrika* 61: 1–17.

REFERENCES

Miller, R. S. 1964a. Interspecies competition in laboratory populations of *Drosophila melanogaster and Drosophila simulans. American Naturalist* 98: 221–238.

Miller, R. S. 1964b. Larval competition in *Drosophila melanogaster and D. simulans. Ecology* 45: 132–148.

Miller, R. S., and J. L. Thomas. 1958. The effects of larval crowding and body size on the longevity of adult *Drosophila melanogaster. Ecology* 39: 118–125.

Milne, A. 1958. Theories of natural control of insect populations. *Cold Spring Harbor Symposium on Quantitative Biology* 22: 253–271.

Møller, H., R. H. Smith and R. M. Sibly, 1989. Evolutionary demography of a bruchia beetle. II. Physiological manipulations. *Functional Ecology* 3: 683–691.

Moore, J. A. 1952a. Competition between *Drosophila melanogaster* and *Drosophila simulans*. I. Population cage experiments. *Evolution* 6: 407–420.

Moore, J. A. 1952b. Competition between *Drosophila melanogaster* and *Drosophila simulans*. II. The improvement of competitive ability through selection. *Proceedings of the National Academy of Sciences USA* 38: 813–817.

Moran, P.A.P. 1953. The statistical analysis of the Canadian lynx cycle. I. Structure and prediction. *Australian Journal of Zoology* 1: 163–173.

Morris, W. F. 1990. Problems in detecting chaotic behavior in natural populations by fitting simple discrete models. *Ecology* 71: 1849–1862.

Morrison, L. D. 1998. The spatiotemporal dynamics of insular ant metapopulations. *Ecology* 79: 1135–1146.

Moss, R., A. Watson, and R. Parr. 1996. Experimental prevention of a population cycle in red grouse. *Ecology* 77: 1512–1530.

Mueller, L. D. 1985. The evolutionary ecology of *Drosophila. Evolutionary Biology* 19: 37–98.

Mueller, L. D. 1987. Evolution of accelerated senescence in laboratory populations of *Drosophila. Proceedings of the National Academy of Sciences USA* 84: 1974–1977.

Mueller, L. D. 1988a. Evolution of competitive ability in *Drosophila* due to density-dependent natural selection. *Proceedings of the National Academy of Sciences USA* 85: 4383–4386.

Mueller, L. D. 1988b. Density-dependent population growth and natural selection in food limited environments: The *Drosophila* model. *American Naturalist* 132: 786–809.

REFERENCES

Mueller, L. D. 1997. Theoretical and empirical examination of density-dependent selection. *Annual Review of Ecology and Systematics* 28: 269–288.

Mueller, L. D., and F. J. Ayala. 1981a. Trade-off between r-selection and K-selection in *Drosophila* populations. *Proceedings of the National Academy of Sciences USA* 78: 1303–1305.

Mueller, L. D., and F. J. Ayala. 1981b. Dynamics of single species population growth: stability or chaos? *Ecology* 62: 1148–1154.

Mueller, L. D. and F. J. Ayala. 1981c. Dynamics of single species population growth: Experimental and statistical analysis. *Theoretical Population Biology* 20: 101–117.

Mueller L. D., and F. J. Ayala. 1981d. Fitness and density dependent population growth in *Drosophila melanogaster*. *Genetics* 97: 667–677.

Mueller, L. D., F. González-Candelas, and V. F. Sweet. 1991a. Components of density-dependent population dynamics: Models and tests with *Drosophila*. *American Naturalist* 137: 457–475.

Mueller, L. D., P. Z. Guo, and F. J. Ayala. 1991b. Density-dependent natural selection and trade-offs in life history traits. *Science* 253: 433–435.

Mueller, L. D., and P. T. Huynh. 1994. Ecological determinants of stability in model populations. *Ecology* 75: 430–437.

Mueller, L. D., A. Joshi, and D. J. Borash. 2000. Does population stability evolve? *Ecology* 81: 1273–1285.

Mueller, L. D., and M. R. Rose. 1996. Evolutionary theory predicts late-life mortality plateaus. *Proceedings of the National Academy of Sciences USA* 93: 15249–15253.

Mueller, L. D., and V. F. Sweet. 1986. Density-dependent natural selection in *Drosophila*: Evolution of pupation height. *Evolution* 40: 1354–1356.

Nachman, G. 1981. Temporal and spatial dynamics of an acarine predator-prey system. *Journal of Animal Ecology* 50: 435–451.

Nachman, G. 1991. An acarine predator-prey metapopulation system inhabiting greenhouse cucumbers. *Biological Journal of the Linnean Society* 42: 285–303.

Naylor, A. F. 1959. An experimental analysis of dispersal in the flour beetle, *Tribolium confusum*. *Ecology* 40: 453–465.

Naylor, A. F. 1965. Dispersal responses of female *Tribolium confusum* to presence of larvae. *Ecology* 46: 341–343.

Nee, S., and R. M. May. 1992. Dynamics of metapopulations: Habitat destruction and competitive coexistence. *Journal of Animal Ecology*, 61: 37–40.

REFERENCES

Nicholson, A. J. 1933. The balance of animal populations. *Journal of Animal Ecology* 2: 131–178.

Nicholson, A. J. 1954a. Compensatory reactions of populations to stresses, and their evolutionary significance. *Australian Journal of Zoology* 2: 1–8.

Nicholson, A. J. 1954b. An outline of the dynamics of animal populations. *Australian Journal of Zoology* 2: 9–65.

Nicholson, A. J. 1957. The self adjustment of populations to change. *Cold Spring Harbor Symposium on Quantitative Biology* 22: 153–173.

Nisbet, R. M., and W.S.C. Gurney. 1982. *Modelling Fluctuating Populations.* John Wiley & Sons, Chichester, UK.

Nunney, L. 1983. Sex differences in larval competition in *Drosophila melanogaster*. The testing of a competition model and its relevance to frequency dependent selection. *American Naturalist* 121: 67–93.

Osborne, K. A., A. Robichon, E. Burgess, S. Butland, R. A. Shaw, A. Coulthard, H. S. Pereira, R. J. Greenspan, and M. B. Sokolowski. 1997. Natural behavior polymorphism due to a cGMP-dependent protein kinase of *Drosophila. Science* 277: 834–836.

Ott, E. 1993. *Chaos in Dynamical Systems.* Cambridge Univ. Press, Cambridge, UK.

Park, T. 1941. The laboratory population as a test of a comprehensive ecological system. *Quartarly Review of Biology* 16: 274–293, 440–461.

Park, T., and M. B. Frank. 1948. The fecundity and development of the flour beetles *Tribolium confusum* and *T. castaneum* at three constant temperatures. *Ecology* 29: 368–374.

Park, T., and M. Lloyd. 1955. Natural selection and the outcome of competition. *American Naturalist* 89: 235–240.

Park, T., D. B. Mertz, and M. Nathanson. 1968. The cannibalism of pupae by adult flour beetles. *Physiological Zoology* 41: 228–253.

Park, T., D. B. Mertz, W. Grodzinski, and T. Prus. 1965. Cannibalistic predation in populations of flour beetles. *Physiological Zoology* 37: 97–162.

Park, T., E. V. Miller, and C. Z. Lutherman. 1939. Studies in population physiology. IX. The effect of imago population density on the duration of the larval and pupal stages in *Tribolium confusum* Duv. *Ecology* 20: 365–373.

Parthasarathy, S., and S. Sinha. 1995. Controlling chaos in unidimensional maps using constant feedback. *Physical Review E* 51: 6239–6242.

303

Pascual, M., and H. Caswell. 1997. Environmental heterogeneity and biological pattern in a chaotic predator-prey system. *Journal of Theoretical Biology* 185: 1–13.

Pearl, R. 1927. The growth of populations. *Quarterly Review of Biology* 2: 532–548.

Pearl, R. 1928. *The Rate of Living.* Knopf, New York.

Pearl, R., and S. L. Parker. 1922. On the influence of density of population upon the rate of reproduction in *Drosophila. Proceedings of the National Academy of Sciences USA* 8: 212–218.

Peters, C. S., M. Mangel, and R. F. Costantino. 1989. Stationary distribution of population size in *Tribolium. Bulletin of Mathematical Biology* 51: 625–638.

Peters, R. H. 1991. *A Critique for Ecology.* Cambridge Univ. Press, Cambridge, UK.

Philippi, T. E., M. P. Carpenter, T. J. Case, and M. E. Gilpin. 1987. *Drosophila* population dynamics: Chaos and extinction. *Ecology* 68: 154–159.

Pimm, S. L., and A. Redfearn. 1988. The variability of population densities. *Nature* 334: 613–614.

Platt, J. R. 1964. Strong inference. *Science* 146: 347–353.

Pray, L. A. 1997. The effect of inbreeding on population-level genetic correlations in the red flour beetle *Tribolium castaneum. Evolution* 51: 614–619.

Press, W. H., B. P. Flannery, S. A. Teukolsky, and W. T. Vetterling. 1986. *Numerical Recipes.* Cambridge Univ. Press, Cambridge, UK.

Prout, T. 1965. The estimation of fitness from population data. *Evolution* 19: 546–551.

Prout, T. 1971a. The relation between fitness components and population prediction in *Drosophila.* I. The estimation of fitness components. *Genetics* 68: 127–149.

Prout, T. 1971b. The relation between fitness components and population prediction in *Drosophila.* II. Population prediction. *Genetics* 68: 151–167.

Prout, T. 1980. Some relationships between density-independent selection and density-dependent population growth. *Evolutionary Biology* 13: 1–68.

Prout, T. 1986. The delayed effect on fertility of preadult competition: Two species' population dynamics. *American Naturalist* 127: 809–818.

Prout, T., and F. McChesney. 1985. Competition among immatures affects their fertility when they grow up: Population dynamics. *American Naturalist* 126: 521–558.

Ranta, E., V. Kaitala, J. Lindström, and E. Helle. 1997b. The Moran effect and synchrony in population dynamics. *Oikos* 78: 136–142.

Ranta, E., V. Kaitala, and P. Lundberg. 1997a. The spatial dimension in population fluctuations. *Science* 278: 1621–1623.

Renshaw, E. 1991. *Modelling Biological Populations in Space and Time*. Cambridge Univ. Press, Cambridge, UK.

Rhodes, O. E. Jr., R. K. Chesser, and M. H. Smith (eds). 1996. *Population Dynamics in Ecological Space and Time*. Univ. of Chicago Press, Chicago, Illinois.

Rich, E. L. 1956. Egg cannibalism and fecundity in *Tribolium*. *Ecology* 37: 109–120.

Richter-Dyn, N., and N. S. Goel. 1972. On the extinction of a colonizing species. *Theoretical Population Biology* 3: 406–433.

Rodriguez, D. J. 1989. A model of population dynamics for the fruit fly *Drosophila melanogaster* with density dependence in more than one life stage and delayed density effects. *Journal of Animal Ecology* 58: 349–365.

Rodriguez, D. J. 1998. Time delays in density dependence are often not destabilizing. *Journal of Theoretical Biology* 191: 95–101.

Rohani, P., R. M. May, and M. P. Hassell. 1996. Metapopulations and equilibrium stability: The effects of spatial structure. *Journal of Theoretical Biology* 181: 97–109.

Rohani, P., and O. Miramontes. 1995. Immigration and persistence of chaos in population models. *Journal of Theoretical Biology* 175: 203–206.

Rohani, P., O. Miramontes, and M. P. Hassell. 1994. Quasiperiodicity and chaos in population models. *Proceedings of the Royal Society of London B* 258: 17–22.

Roland, J., and P. D. Taylor. 1995. Herbivore-enemy interactions in fragmented and continuous forests. In *Population Dynamics: New Approaches and Synthesis*, ed. N. Cappuccino and P. W. Price, 195–208. Academic Press, San Diego, California.

Rorres, C. 1979. Local stability of a population with density-dependent fertility. *Theoretical Population Biology* 16: 283–300.

Rose, M. R. 1984. Laboratory evolution of postponed senescence in *Drosophila melanogaster*. *Evolution* 38: 1004–1010.

Rose, M. R. 1991. *Evolutionary Biology of Aging*. Oxford Univ. Press, New York.

Rose, M. R., T. J. Nusbaum, and A. K. Chippindale. 1996. Laboratory evolution: The experimental wonderland and the Cheshire cat syndrome. In *Adaptation*, ed. M. R. Rose and G. V. Lauder, 221–241. Academic Press, San Diego, California.

Roughgarden, J. 1971. Density-dependent natural selection. *Ecology* 52: 453–468.

Roughgarden, J. 1975. A simple model for population dynamics in stochastic environments. *American Naturalist* 109: 713–736.

Roughgarden, J. 1976. Resource partitioning among competing species—a coevolutionary approach. *Theoretical Population Biology* 9: 388–424.

Roughgarden, J. 1979. *Theory of Population Genetics and Evolutionary Ecology: An Introduction.* MacMillan, New York.

Royama, T. 1971. A comparative study of models for predation and parasitism. *Researches on Population Ecology* Supp. 1: 1–91.

Royama, T. 1977. Population persistence and density dependence. *Ecological Monographs* 47: 1–35.

Royama, T. 1992. *Analytical Population Dynamics.* Chapman & Hall, London.

Ruxton, G. D. 1994. Low levels of immigration between chaotic populations can reduce system extinctions by inducing asynchronous regular cycles. *Proceedings of the Royal Society of London B* 256: 189–193.

Ruxton, G. D. 1996a. Dispersal and chaos in spatially structured models: An individual approach. *Journal of Animal Ecology* 65: 161–169.

Ruxton, G. D. 1996b. Synchronization between individuals and the dynamics of linked populations. *Journal of Theoretical Biology* 183: 47–54.

Ruxton, G. D., and P. Rohani. 1998. Population floors and the persistence of chaos in ecological models. *Theoretical Population Biology* 53: 175–183.

Sabelis, M. W., O. Diekmann, and V.A.A. Jansen. 1991. Metapopulation persistence despite local extinction: Predator-prey patch models of the Lotka-Volterra type. *Biological Journal of the Linnean Society* 42: 267–283.

Sang, J. H. 1949. Population growth in *Drosophila* cultures. *Biological Review* 25: 188–219.

Santos, M., D. J. Borash, A. Joshi, N. Bounlutay, and L. D. Mueller. 1997. Density-dependent natural selection in *Drosophila*: Evolution of growth rate and body size. *Evolution* 51: 420–432.

Schaffer, W. M. 1984. Stretching and folding in lynx fur returns: Evidence for a strange attractor in nature? *American Naturalist* 124: 798–820.

Sheeba, V., and A. Joshi. 1998. A test of simple models of population growth using data from very small populations of *Drosophila melanogaster. Current Science* 75: 1406–1410.

REFERENCES

Shiotsugu, J., A. M. Leroi, H. Yashiro, M. R. Rose, and L. D. Mueller. 1997. The symmetry of correlated responses in adaptive evolution: An experimental study using *Drosophila*. *Evolution* 51: 163–172.

Sinha, S., and S. Parthasarathy. 1994. Behaviour of simple population models under ecological processes. *Journal of Biosciences* 19: 247–254.

Sinha, S., and S. Parthasarathy. 1996. Unusual dynamics of extinction in a simple ecological model. *Proceedings of the National Academy of Sciences USA* 93: 1504–1508.

Sokal, R. R., and F. J. Rohlf. 1981. *Biometry*. 2nd ed. W. H. Freeman and Co., New York.

Sokoloff, A. 1972. *The Biology of* Tribolium. Vol. 1. Oxford Univ. Press, Oxford, UK.

Sokoloff, A. 1974. *The Biology of* Tribolium. Vol. 2. Oxford Univ. Press, Oxford, UK.

Sokoloff, A. 1977. *The Biology of* Tribolium. Vol. 3. Oxford Univ. Press, Oxford, UK.

Sokolowski, M. B. 1980. Foraging strategies of *Drosophila melanogaster*: A chromosomal analysis. *Behavior Genetics* 10: 291–302.

Sokolowski, M. B., H. S. Pereira, and K. Hughes. 1997. Evolution of foraging behavior in *Drosophila* by density-dependent selection. *Proceedings of the National Academy of Sciences USA* 94: 7373–7377.

Solé, R. V., and J. Valls. 1992. On structural stability and chaos in biological systems. *Journal of Theoretical Biology* 155: 87–102.

Soulé, M. E., and D. Simberloff. 1986. What do genetics and ecology tell us about the design of nature reserves? *Biological Conservation* 35: 18–40.

Spencer, H. 1864. *The Principles of Biology*. Appleton, New York.

Spencer, M., and P. H. Warren. 1996. The effects of habitat size and productivity on food web structure in small aquatic microcosms. *Oikos* 75: 419–430.

Stacey, P. B., V. A. Johnson, and M. L. Taper. 1997. Migration within metapopulations: The impact upon local dynamics. In *Metapopulation Biology: Ecology, Genetics, and Evolution*, ed. I. Hanski and M. E. Gilpin, 267–291. Academic Press, San Diego, California.

Steele, J. H. 1985. A comparison of terrestrial and marine ecological systems. *Nature* 313: 355–358.

Stokes, T. K., W.S.C. Gurney, R. M. Nisbet, and S. P. Blythe. 1988. Parameter evolution in a laboratory insect population. *Theoretical Population Biology* 34: 248–265.

REFERENCES

Strong, D. 1986. Density-vague population change. *Trends in Ecology and Evolution* 1: 39–42.

Sugihara, G. 1995. From out of the blue. *Nature* 378: 559–560.

Sugihara, G., and R. M. May. 1990. Nonlinear forecasting as a way of distinguishing chaos from measurement error in time series. *Nature* 344: 734–741.

Sulzbach, D. S., and J. M. Emlen. 1979. Evolution in competitive ability in mixtures of *Drosophila melanogaster*. Populations with an initial asymmetry. *Evolution* 33: 1138–1149.

Swick, K. E. 1981. Stability and bifurcation in age-dependent population dynamics. *Theoretical Population Biology* 20: 80–100.

Tatar, M., and J. R. Carey. 1994. Genetics of mortality in the bean beetle *Callosobruchus maculatus*. *Evolution* 48: 1371–1376.

Tatar, M., and J. R. Carey. 1995. Nutrition mediates reproductive trade-offs with age-specific mortality in the beetle *Callosobruchus maculatus*. *Ecology* 76: 2066–2073.

Tatar, M., J. R. Carey, and J. W. Vaupel. 1993. Long-term cost of reproduction with and without accelerated senescence in *Callosobruchus maculatus*: Analysis of age-specific mortality. *Evolution* 47: 1302–1312.

Thomas, W. R., M. J. Pomerantz, and M. E. Gilpin. 1980. Chaos, asymmetric growth and group selection for dynamical stability. *Ecology* 61: 1312–1320.

Tilman, D., and D. Wedin. 1991. Oscillations and chaos in the dynamics of a perennial grass. *Nature* 353: 653–655.

Travisano, M., J. A. Mongold, A. F. Bennett, and R. E. Lenski. 1995a. Experimental tests of the roles of adaptation, chance and history in evolution. *Science* 267: 87–90.

Travisano, M., F. Vasi, and R. E. Lenski, 1995b. Long-term experimental evolution in *Escherichia coli*. III. Variation among replicate populations in correlated responses to novel environments. *Evolution* 49: 189–200.

Tucic, N., D. Cvetkovic, V. Stojilkovic and D. Bejakovic. 1990. The effects of selection for early and late reproduction on fecundity and longevity in bean weevil *(Acanthoscelides obtectus)*. *Genetica* 80: 221–227.

Tucic, N., I. Gliksman, D. Šešlija, D. Milanovic, S. Mikuljanac, and O. Stojkovic, 1996. Laboratory evolution of longevity in the bean weevil *(Acanthoscelides obtectus)*: The effects of density-dependent and age-specific selection. *Journal of Evolutionary Biology*, 9: 485–503.

REFERENCES

Tucic, N., O. Stojkovic, I. Gliksman, D. Milanovic, and D. Šešlija, 1997. Laboratory evolution of life-history traits in the bean weevil (Acanthoscelides obtectus): The effects of density-dependent and age-specific selection. *Evolution* 51: 1896–1909.

Turchin, P. 1990. Rarity of density dependence or population regulation with time lags? *Nature* 344: 660–663.

Turchin, P. 1991. Reconstructing endogenous dynamics of a laboratory *Drosophila population. Journal of Animal Ecology* 60: 1091–1098.

Turchin, P. 1993. Chaos and stability in rodent population dynamics: Evidence from nonlinear time-series analysis. *Oikos* 68: 167–172.

Turchin, P. 1995a. Population regulation: Old arguments and a new synthesis. In *Population Dynamics: New Approaches and Synthesis*, ed. N. Cappuccino and P. W. Price, 19–40. Academic Press, San Diego, California.

Turchin, P. 1995b. Chaos in microtine populations. *Proceedings of the Royal Society of London B* 262: 357–361.

Turchin, P., and I. Hanski. 1997. An empirically based model for latitudinal gradient in vole population dynamics. *American Naturalist* 149: 842–874.

Turchin, P., and A. D. Taylor. 1992. Complex dynamics in ecological time series. *Ecology* 73: 289–305.

Turelli, M. 1978. A reexamination of stability in randomly varying versus deterministic environments with comments on the stochastic theory of limiting similarity. *Theoretical Population Biology* 13: 244–267.

Turelli, M., and D. Petry. 1980. Density-dependent selection in a random environment: An evolutionary process that can maintain stable population dynamics. *Proceedings of the National Academy of Sciences USA* 77: 7501–7505.

Uvarov, B. P. 1931. Insects and climate. *Transactions of the Entomological Society of London* 79: 1–249.

Van de Klashorst, G., J. L. Readshaw, M. W. Sabelis, and R. Lingemann. 1992. A demonstration of local asynchronous cycles in an acarine predator-prey system. *Exp. Appl. Acar.* 14: 185–199.

Van der Meijden, E., and C.A.M. van der Veen-van Wijk. 1997. Tritrophic metapopulation dynamics: A case study of ragwort, the cinnabar moth and the parasitoid *Cotesia popularis*. In *Metapopulation Biology: Ecology, Genetics, and Evolution*, ed. I. Hanski and M. E. Gilpin, 387–405. Academic Press, San Diego, California.

309

Van der Meijden, E., C.A.M. van der Veen-van Wijk, and R. E. Kooi. 1991. Population dynamics of the cinnabar moth (*Tyria jacobaeae*): Oscillations due to food limitation and local extinction risks. *Neth. Jour. Zool.* 41: 158–173.

Vasi, F., M. Travisano, and R. E. Lenski. 1994. Long-term experimental evolution in *Escherichia coli*. II. Changes in life-history traits during adaptation to a seasonal environment. *American Naturalist*, 144: 432–456.

Wade, M. J. 1990. Genotype-environment interaction for climate and competition in a natural population of flour beetles *Tribolium castaneum*. *Evolution* 44: 2004–2011.

Walde, S. J. 1995. Internal dynamics and metapopulations: Experimental tests with predator-prey systems. In *Population Dynamics: New Approaches and Synthesis*, ed. N. Cappuccino and P. W. Price, 173–193. Academic Press, San Diego, California.

Walls, R. C., and D. L. Weeks. 1969. A note on the variance of a predicted response in regression. *American Statistician* 23: 24–26.

White, A., M. Begon, and R. G. Bowers. 1996. Explaining the colour of power spectra in chaotic ecological models. *Proceedings of the Royal Society of London B* 263: 1731–1737.

Williams, J. 1985. Statistical analysis of fluctuations in Red Grouse bag data. *Oecologia* (Berlin) 65: 269–272.

Worthen, W. B. 1989. Predator-mediated coexistence in laboratory communities of mycophagous *Drosophila* (Diptera: Drosophilidae). *Ecological Entomology* 14: 117–126.

Worthen, W. B., and J. L. Moore, 1991. Higher-order interactions and indirect effects—a resolution using laboratory *Drosophila* communities. *American Naturalist* 138: 1092–1104.

Wright, S. 1931. Evolution in Mendelian populations. *Genetics* 16: 97–159.

Wright, S. 1940. The statistical consequences of Mendelian heredity in relation to speciation. In *The New Systematics*, ed. J. S. Huxley, 161–183. Clarendon Press, Oxford, UK.

Young, A. M. 1970. Predation and abundance in populations of flour beetles. *Ecology* 51: 602–619.

Author Index

Subject Index